SCIENCE, TECHNOLOGY, AND SOCIETY

New Directions

Andrew Webster

Rutgers University Press
New Brunswick, New Jersey

First published in cloth and paperback in the United States, 1991,
by Rutgers University Press

First published in cloth and paperback in the United Kingdom, 1991,
by Macmillan Education, Ltd.

Copyright © 1991 by Andrew Webster

Printed in Hong Kong

Library of Congress Cataloging-in-Publication Data
Webster, Andrew
Science, technology, and society: new directions
1. Science, Sociopolitical aspects. 2. Technology, Social aspects.
I. Title.
ISBN 0-8135-1722-2 (cloth)   ISBN 0-8135-1723-0 (pbk.)

Contents

Foreword

This book charts recent developments in the sociology of science as well as changes in the character of science and technology as social institutions. The idea for the book grew out of my belief that a text which linked the insights of sociology with the concerns of science policy was needed and might be especially useful for students on both natural and social science courses pursuing the broad dynamics of the relationship between science and society.

The 'New Directions' mapped out here could not have been so without some 'old' academic friends who have contributed over a number of years to my own understanding of the field. Many of the issues raised in the text have been discussed more or less directly with Henry Etzkowitz, Mike Gibbons, Paul Hoch, Mike Mulkay, Paul Quintas, Nick Read, Joyce Tait, and colleagues at the Science Policy Support Group, Peter Healey and John Ziman.

But I especially want to thank Wendy Faulkner, Steve Yearley and Julian Constable for reading and commenting on parts of the manuscript and helping me to refine and elaborate on the argument. None of these is responsible, of course, for any 'mistaken' directions that the text takes. I should also like to thank Dilys Jones, commissioning editor, and Keith Povey, editorial services, for their help in producing the text with both speed and efficiency.

The very valuable support of colleagues has been matched, however, by that of my wife, Helen, whose patience and good humour made the writing of this book not only a possibility but a reality that was much more enjoyable than should be allowed.

ANDREW WEBSTER

Acknowledgements

The author and publishers wish to thank the following who have kindly given permission for the use of copyright material:

BSA Publications Ltd for a table from 'Inventions from R&D', by F. Hull, *Sociology*, vol. 22, no. 3, August 1988.

Longman Group Ltd for a diagram from H. Rothman, 'Science Mapping and Strategic Planning', in M. Gibbons *et al.* (eds), *Science and Technology Policy in the 1980s and Beyond* (1984).

Penguin Books Ltd for use of a diagram from *The Double Helix* by J. D. Watson (1970).

Sage Publications Ltd for material from J. Fujimura, 'Constructing Do-able Problems in Cancer Research', *Social Studies of Science*, vol. 17, pp. 257–93; and P. Vergragt, 'The Social Shaping of Industrial Innovations', *Social Studies of Science*, vol. 18, pp. 303–60.

Frances Pinter Publishers Ltd for a chart from E. Braun, *Wayward Technology* (1984).

List of Abbreviations

ABRC	Advisory Board for the Research Councils
ACOST	Advisory Committee on Science and Technology
ALF	Animal Liberation Front
AT	Alternative Technology
BRIDGE	Biotechnology Research for Innovation
BTG	British Technology Group
CAITS	Centre for Alternative Industrial and Technological Systems
CEST	Centre for the Exploitation of Science and Technology
CIMAH	Control of Industrial Major Hazards
DA	Discourse Analysis
DNA	Deoxyribonucleic Acid
EDA	Ethnography and Discourse Analysis
GASP	Generic Analytical Sample Preparation
IRC	Interdisciplinary Research Centre
ITDG	Intermediate Technology Development Group
MAST	Marine Science and Technology
MITI	Ministry of International Trade and Industry (Japan)
MNC	Multinational Corporation
N/C	Numerical Control
SAST	Strategic Analysis in Science and Technology
SCOT	Social Construction of Technology
TEA	Transversely Excited Atmospheric (Laser)

1 Introduction

The 'Discovery Dome' has displaced the glass encased, static exhibits. Visitors to the state-of-the art Science Museum are no longer simply asked to admire the scientific and technological wonders of the world: they must be encouraged to understand the principles that lie behind them in a way that is simple, entertaining and 'hands-on'.

We should welcome this attempt to increase our awareness of the enormous importance that science and technology play in our daily lives. But one wonders whether much has really changed. A prior understanding of the how, when and why of science and technology could be considered a vital part of anyone's education. But, it is likely that, for many, the understanding of science derived from the 'Discovery Dome' is partial, fleeting and as difficult to relate to daily life as it ever was. Despite the 'user-friendly' style, at the heart of the interactive museum lies the conventional image of science as *asocial, non-political, expert and progressive*. Like their more solemn and aloof predecessors, modern museums typically construct an image of science that conveys a sense of power and authority.

Indeed, the authority of science relies precisely on our perceiving it as something that lies outside society: science is not, in these museums, a contested terrain, an arena where differences of opinion and division appear. The very fact that science has enabled 'us' to land on the Moon, to produce polyethylene plastics, to transplant organs, to engineer genes and so on requires us to marvel at its utlity and admire its certainty. Science deals with 'facts'.

Outside the museum, however, science and associated technologies can take on a rather different character. The image of science may be much more negative; it may appear more distant, remote, less under our control and even harmful to our very survival. No museum exhibit displays the *collapse* of a technological system, a breakdown which may indeed be fatal as in the Zeebrugge Ferry and Chernobyl power station disasters of recent years. At such times, the polished, professional and progressive image of science becomes tarnished and the public comes to realise just how much trust it has to invest in complex technological systems the dismantling of which is often not easy. A decade after the partial meltdown of the nuclear reactor at Three Mile Island in Pennsylvania, the process of cleaning up the radioactive debris continues at a cost of over $1 000 million, three times as much as it cost to build the plant in the first place.

In some countries the anxiety and distrust about some technologies has led to what one American scientist has called 'science-bashing': Donald Kennedy, President of Stanford University in California recently criticised what he saw as a 'new and corrosive popular mistrust of scientists and their work'. Kennedy urged his fellow scientists to show 'more stiffness in the face of the special political interests that are hostile to American science' (1989). Science has become politicised and must in the view of many be made to be accountable: its effects are challenged by environmentalist pressure groups such as Jerry Rifkin's Economic Foundation in the US and the UK's Greenpeace movement. The reverential respect for scientific facts is replaced by 'technodread'.

What is going on here? Why such divergent views towards science and technology? To make any sense of this situation will require an analysis that recognises from the start that science and technology are rooted in the society which creates them. Just as the Discovery Dome seeks to create an image of science that is authoritative, so the pressure groups tell of its authoritarian nature. The nature of 'science' is then very much a matter of debate within society and the outcome of that continuing debate reproduces or changes the social institution of what we understand as 'science'.

This book provides an introduction to some of the key dimensions of science and technology as social processes and the debates about their impact on society. It draws on a range of sociological analyses that have explored the culture of scientific research, the relations between science and technology, their controversial effects and the increasing attempts by government in industrialised countries to regulate them.

But the book also claims to explore 'new directions' in the relationship between science, technology and society. What might these be? While these will be debated more fully throughout the book, a preliminary indication of them can be given here.

(i) *First*, the very nature of scientific inquiry within the laboratory is changing precisely because experiments tend to be the work of teams of scientists rather than isolated individuals beavering away at a laboratory bench. These teams rely on increasingly costly and complex instruments and apparatus and, most importantly, require collaboration across a range of disciplines. This has had two fairly obvious effects: first disciplinary boundaries have become increasingly blurred, especially in the 'new' sciences, such as biotechnology, materials science and information technology. Secondly, the distinction between basic and applied research begins to look less and less useful as the research team's efforts pull in both directions at the same time, indeed as a recent report of the UK's Advisory Board for the Research Councils said, this distinction was 'a meaningless dichotomy' in a growing number of research fields (ABRC, 1986). The research environment may reflect this: academic departments are increasingly being restructured to accommodate basic and applied interests, as the example from Imperial College, London suggests below.

What's this: 'pure' or 'applied'?

Dr Andrew Goldsworthy of the Department of Pure and Applied Biology first became interested in the role of electric currents in plant development when he detected and measured minute currents, hundreds of thousands of times smaller than those which light a torch, passing through plants . . . He is now using a 'vibrating probe' to measure the strength and direction of the currents flowing through cells . . .

Goldworthy's success in increasing the rate of regeneration of shoots from tobacco calluses and, recently, the rate of regeneration of wheat plants from tissue cultures suggests that, for two important crops at least and probably for others, the very cheap technique of very low level electrical stimulation might make the difference between commercial success or failure for genetic engineering technology.

Source: *Biobulletin* (SERC, October 1987).

(ii) At a more general level, the distinction between 'science' and 'technology' also begins to lose its relevance *in practice* even if in principle it is still possible to distinguish the two. In practice, the advancement of science and the advancement of technology may depend on investigators in certain fields asking very similar questions. Rather than trying to distinguish 'pure' technologists from 'pure' scientists in terms of the character of research they undertake, we should understand their differences in terms of the social context or circumstances within which they work. One might expect, for example, to find people calling themselves 'scientists' working within academic research labs while 'technologists' appear in industry. But even this isn't a very meaningful or helpful distinction since more *practitioners* of science, whether 'pure' or 'applied' are actually found within *industry* than anywhere else. Science *and* technology are very much industrialised today.

One can see this in a variety of ways. For example, some of the more recent Nobel Prize winners have been based in industrial research labs. But perhaps the most obvious way this is apparent is simply in terms of the amount of R&D that is conducted within industry by company research staff, especially in the chemical and electronic industries (see Table 1.1).

Table 1.1   In-house expenditure on R&D in broad industry sectors (£m)

Sector	1987
Chemicals industries	1303.0
Mechanical engineering	285.6
Electronics	1854.7
Other electrical engineering	142.2
Motor vehicles	450.5
Aerospace	870.9
Other	462.8

(iii) *Thirdly*, the existence of corporate labs conducting basic research is not new. Many of today's biggest companies grew by virtue of the commercialisation of products discovered and developed within their own in-house labs. This process has been explored by historians of science such as Dennis (1987) and Hounshell (1988). But today, this reliance on corporate innovation is much more significant for the major companies of industrialised capitalist economies (see Freeman, 1974): while mergers may be one effective way of buying out possible or actual competitors, without new products to sustain and expand market share, companies grow moribund and lack direction. The exploitation of scientific knowledge is the key to survival for the biggest multinational corporations, as well as many small new 'high-tech' firms.

(iv) Finally, science and technology are now firmly located within the political arena because of their central concern to the state: all governments today are involved in the large-scale funding, management and regulation of science and technology. There is, however, considerable variation in the strategies and priorities that different governments adopt towards science. Some, such as France, the Netherlands and Japan, have developed fairly strong direction from the centre whereas others, such as the UK and USA, have more pluralistic, some might argue disjointed, policies towards their science.

At the same time, the increasing cost of what Price (1984) has called 'Big Science' has led to the development of expensive research centres dependent on national and increasingly international state funding. In the United States, for example, the federal government recently approved the establishment in Texas of a $4.4 billion superconducting supercollider, a 53-mile particle accelerator. Within Europe, there is being established an increasing number of international research programmes between member states of the European Community. Common to these state and inter-state initiatives is the attempt to foster improved processes for the transfer of technology from the laboratory – whether in academia or industry – to the marketplace.

One important overall effect of these four emergent trends is that the traditional view that science and technology are 'neutral', uninfluenced by wider social processes, can no longer be sustained.

As we shall see shortly, sociologists have long recognised this anyway: it is now something increasingly acknowledged by scientists themselves. Aronowitz (1989) has argued that science and technology lose this sense of 'neutrality' the more they are tied to the priorities of capitalist industry and the state, a process apparent in the four trends identified above. This connection makes science in some sense more powerful but at the same time makes its claims inherently more contestable as it loses its image of autonomy. It seems increasingly difficult for scientists to present themselves as independent experts on matters of social interest whether these be about nuclear power, genetically engineered organisms, food hygiene, toxic waste or whatever.

These shifts in the character of science and technology could be regarded as 'natural' developments of earlier processes, as being in some way simply evolutionary trends from a previous stage. However, it is the simultaneous combination of all four processes that suggests the possible emergence of a qualitatively new character to the social institution we call science. Recent empirical research by sociologists, which will be discussed later, seems to provide some warrant for this. The sociological analysis of science, however, has a fairly long pedigree, and to understand recent debates we need to examine, if only briefly, the earlier work in the field.

Developments in the sociology of science

For much of the development of the discipline, sociology has regarded scientific knowledge as a special domain free from social influence, a view still typically held by many non-sociologists today. The classical sociologists such as Comte, Durkheim, Marx and Weber, believed that science was an independent, objective form of knowledge furthering the development of society. Sociology itself could endeavour to be 'scientific' in its analysis of this development.

When these and other sociologists explored the historical and cultural roots of knowledge and belief – particularly important for the early social anthropologists – scientific knowledge was traditionally excluded from the debate. The sociology of knowledge should concern itself with religion, magic, great historical myths and social ideologies but not with science, which was seen as 'outside' society.

This view prevailed for many years and informed what are usually regarded as the first attempts to construct a *sociology of science* proper. It was the work of the American functionalist sociologist, Robert Merton, in the 1950s that first encouraged an exploration of science as a social institution, though he too excluded from consideration scientific knowledge *per se*. Merton initiated an inquiry into the organisational and behavioural aspects of the scientific enterprise, which was then extended through the work of his fellow functionalists, Storer and Hagstrom.

Merton (1949) claimed that the institution of science operates according to a set of prescribed *norms* that together make up a broad 'ethos' to which scientists are – or at least should be – committed. These are the rules of the scientific game, directing scientists to act in ways which will sustain the ideal organisation and functioning of science. The four principal norms are universalism, communality, organised scepticism and disinterestedness. Simply put, scientists should with caution but with an open mind treat any knowledge-claim on its merits. They should share with their peers, in a free exchange, ideas that are subject to close scrutiny and free from personal interest or ambition. For Merton, the history of science shows how conformity to these institutional codes of practice helps to ensure the production of objective scientific knowledge.

One difficulty with Merton's thesis was showing *why* scientists should behave in this way: for Merton, it was because of their 'deep devotion to the advancement of knowledge'. But this still begs the question of the origins of this devotion. Storer (1966) believed that such devotion derives from scientists' desire for professional recognition, the 'competent response' of other scientists to their creativity. To gain the respect and support of their peers, scientists act in such a way that the overall objectivity and warrant of scientific knowledge is sustained.

Though the first and in many ways highly sophisticated sociological account of science, Merton's thesis was soon to come under attack from a number of sociologists who, through more detailed empirical work than the functionalists had ever done, began to challenge the Mertonian conception of a 'scientific community' conforming to a 'scientific ethos'. Mulkay (1976) was one of the first to provide an empirically-based critique of the alleged institutionalisation of the 'norms' identified by Merton. He showed how it is more plausible to interpret 'norms' as a repertoire of claims that scientists use quite variably, as contexts change, in order to justify their particular behaviour. In some circumstances, then, a scientist might challenge another scientist for failing to conform, say, to the rule of disinterestedness, perhaps because of witholding some information about an experiment. Such a critic might, on the other hand just as easily resort to similar tactics and justify these on the grounds that 'further research' within his or her own research team was necessary before more data were released. In turn, other scientists might accuse the team of contravening the norm of disinterestedness, and so on. This was well illustrated recently in the argument and counter-argument about the release of information concerning the Pons/Fleischmann experiments in 1989 on nuclear fusion. Mulkay argues therefore that the successful use of normative *claims* to justify behaviour help to protect particular group interests in science, and more generally, protect the collective interests of science from 'outside interference'. In short, they constitute what Mulkay calls the 'occupational ideology' of science: accordingly, he argues, 'the original functionalist analysis did identify a genuine social reality, but one better conceived as an ideology than a normative structure' (1976b, p.465).

So how then are scientists rewarded for their work, if this is not via prescribed institutional norms of recognition, as claimed by Merton? Mulkay argues that 'Researchers are simply rewarded for communicating information which their colleagues deem to be useful in the pursuit of their own studies' (1976, p.642). What is or is not 'useful' will vary according to context, including the long- and short-term interests of researchers, their financial circumstances, the attitude of the government and commercial institutions towards the new ideas, and so on. There is no simple meritocracy operating in science. Groups that are well-placed in the scientific establishment are likely to receive greater support for their work. This is a view that has also been shared by the French sociologist, Bourdieu, who writing from within a neo-Marxist perspective, has claimed that 'The market in scientific goods has its laws and they are nothing to do with ethics or norms . . . . Claims to legitimacy [of ideas] draw their legitimacy from the relative strength of the group whose interests they express' (1975, p.40). This implies that some scientists and their research areas may enjoy a favourable position with regard to the receipt of recognition which is *independent* of the 'quality' of the work they produce. There is in fact now available a good deal of sociological evidence that supports this view. For example, Wynne (1979) showed how the Edinburgh physicist, C. G. Barkla, could draw on his 'cultural capital' as Professor and one-time Nobel Prize winner to ensure continued recognition and even reward for his work on what he called the 'J Phenomenon' despite this same work being held to be inconsistent with the prevailing scientific orthodoxy and being widely discredited.

Challenging the image of science

These developing critiques, both theoretical and empirical, of the Mertonian account of science marked a shift away from the traditional belief that science was in some way 'outside' society: they not only challenged accounts of scientists' behaviour but also demanded that scientific *ideas* themselves must be subject to sociological analysis. This analysis had begun to show how the meaning and status of scientific knowledge is rarely stable across groups of scientists and that the attainment of a consensus among scientists usually depends on their sharing similar professional interests. I shall give some detailed examples of this work in Chapter 2. While this new perspective was emerging through a growing volume of empirical work among sociologists during the 1970s, similar ideas were being developed outside sociology by radical historians and philosophers of science.

From within the radical tradition there emerged during the 1970s a growing number of European and British neo-Marxists who sought to show how the practice and claims of science are influenced by the social and material interests of capitalism. Much of this work echoed, while going much further than, the earlier, now considered 'classic', text of Boris Hessen (1931), the Soviet historian, who had argued that Newton's laws of physics suited the interests of early capitalism. Woesler (1976) argued that just as the technical division of labour within industry became increasingly specialised so too did the pursuit of knowledge within science which itself becomes ever more closely tied to industrial interests. A number of Italian Marxists, such as Ciccotti (1976), argued that scientific knowledge serves the material needs of capitalism: so, for example, the shift from a 'mechanistic' physics in the early twentieth century to a non-reductionist particle physics today is said to 'correspond with the passage from mechanisation to automation in the productive process for material goods' (p.50).

Similar ideas were developed by members of the 'Radical Science' movement in Britain, such as Hilary and Steven Rose (1976). A growing number of empirically-based critiques of 'capitalist science and technology' appeared with a strong emphasis on the alienating and destructive role of science in the workplace (Levidow, 1982; Gorz, 1976), in medicine (Navarro, 1986, Epstein, 1979), and in the wider environment (Carson, 1965).

The notion that the content of science could be regarded as being socially grounded was also to find expression – but completely independently of the neo-Marxists – in the work of a number of philosopher-historians of science, most notably the American, Thomas Kuhn. He argued that the 'facts' and accepted theories of science were the result of a continual process of negotiation among scientists. At first sight this may not seem earth-shattering news: but the vital point for Kuhn was that the criteria scientists used to evaluate competing knowledge claims as part of this negotiation were very far removed from the methodological canons of scientific debate laid down by philosophers of science: Popper (1972) for example, had argued that only those scientific claims that had successfully run the gauntlet of attempts by scientists to *falsify* them were to be accepted as true.

Kuhn's basic thesis, however, was that scientific ideas were acceptable so long as they conformed broadly to the existing 'paradigm' or model of scientific understanding operating within and, in the more powerful paradigms between, disciplines. Overarching paradigms provide the orthodox wisdom of any one area of science, to some degree acting as the 'dogma' of institutionalised science. In this sense scientists are conformist and conservative. Scientists do not go out to falsify others' claims nor to discover the 'unknown' but to uncover the 'known', that is, what the paradigm tells them to expect from their experiments in the lab. Scientists know what they are looking for in advance, and anticipate results, discarding those that don't 'fit' as either artefacts of the experiment or unexplained 'anomalies' that can be conveniently ignored. Only when the weight of such anomalies becomes too great to ignore does pressure build up for a change in the paradigmatic framework.

How then do *new* ideas ever get accepted? For Kuhn this depends on the persuasive power of scientists to convince others of their findings, noting that those in elite positions within science may well be best placed to do this though others in similar positions might as readily resist such a threat to the established paradigm. There is no sign here of the Mertonian norms of science acting as institutional constraints on scientists: instead, scientists appear to choose those ideas which conform to their image of 'good science'. This may vary according to their personal subscription to one or more of a set of values (such as 'simplicity', 'comprehensiveness', 'utility' and so on) which they use to judge the relative merits of different claims. None of these evaluative judgements nor the way they are applied can be said to conform to some asocial pre-given standard of rationality. 'Scientific worth' immediately becomes socially grounded and relative. As Dolby (1972) has argued:

> There can only be scientific knowledge of what a group of people can agree upon. This . . . introduces the possibility of relativism in the standards of those to whom a scientist directs his arguments. In considering objectivity in science, one must consider the audience for a knowledge-claim. (p.317).

'Science' as a social construction

These developments in the history and philosophy of science were paralleled by a rapid growth of sociological research on science during the 1970s that sought to understand the social basis not merely of scientists' behaviour but also of their ideas. In part, this work took its lead from Kuhn, relishing the new opportunity to explore the social context within which scientific knowledge became certified. A common feature of this work was a focus on controversy in science, the resistance to innovation and the struggle between orthodox and heterodox groups in science competing for scientific reward.

The study of controversy was seen to offer a particularly useful insight into the way in which the status of scientific knowledge was dependent on negotiation and debate between interested parties, both the practitioners of science and 'lay' people. Such controversies throw light, for example, on the way scientific 'expertise' could in certain circumstances be regarded by participants as highly contentious: one group's expert was another's fool. In such debates, the very nature of what counts as 'good evidence' is itself a matter of controversy. Sociologists could see how scientists in different specialties responded to this challenge to their 'authority', a response that typically involves a whole host of social strategies and rhetorics which bear no resemblance to the idealised principles of the 'scientific method'.

Nelkin (1979) has provided a rich variety of empirical research illustrating the social aspects of scientific controversy. She shows, for example, how publicly controversial issues such as the development of a new drug, the siting of nuclear power plants or the decision to expand an airport involve participants in claims and counter-claims that cannot be easily resolved by an appeal to the 'evidence', since there is no necessary consensus on what the 'facts' are. As she says,

> [I]n all disputes broad areas of uncertainty are open to conflicting scientific interpretation. Decisions are often made in a context of limited knowledge about potential social or environmental impacts and there is seldom conclusive evidence to reach definitive resolution. Thus power hinges on the ability to manipulate knowledge to challenge the evidence presented to support particular policies, and technical expertise becomes a resource exploited by all parties to justify their political and economic views. (p.16)

It is for these very reasons that it is not uncommon to find drugs, pesticides or whatever banned in some countries but not others (see Gillespie *et al.* 1979, discussed in Chapter 5).

Similar themes were developed during the 1970s in a growing body of work that focused on deviant knowledge, knowledge-claims which in some way failed to conform to the social and scientific 'paradigm' of orthodoxy. Studies on parapsychology (Collins and Pinch, 1979) and astrology (Wright, 1979) are two good examples. The author's own study on the debate over acupuncture in the early 1970s also exemplifies this type of work (Webster, 1979; 1981). In this case it was shown that orthodox medics in the US and in Europe used a variety of occupational strategies to contain the emergence of the Chinese treatment at a time of growing popular interest in alternative medicine. These strategies had little to do with any open, 'disinterested' examination of research being conducted in both China and the West on the mechanisms behind acupuncture's painkilling powers. Only when the non-orthodox, traditional acupuncturists had been controlled legally by the courts (or Parliament) and the language of their practice had been translated into a discourse that could be acceptable to the medic did this hostility subside: acupuncture became 'transcutaneous nerve stimulation'.

Other work has focused on controversial knowledge claims made *from within*, by members of the scientific community itself. Here the work of Collins (1981, 1985) is of particular importance, providing further insight into the socially-negotiated nature of what is to be regarded as acceptable scientific knowledge. One of Collins's early studies (1975) examined a dispute within physics over the possible existence of 'gravitational waves', a form of cosmic radiation predicted by Einsteinian theory, resulting from massive astrophysical events such as the collapse of a star. Most physicists accepted that such waves should exist and agreed broadly on how one would go about the very difficult, delicate task of finding and measuring them. Nevertheless, there was considerable uproar when one physicist, after many years' effort, actually claimed to have found them. Many dismissed the finding as the result of poor experimentation or even fabrication of data. Participants to the controversy endeavoured to establish their own ground-rules for what a 'good, competent' experiment on gravitational waves might involve: but since there was no overall consensus on this, any experimental technique could be challenged and any 'evidence' it produced was regarded as ambiguous. Collins shows how the accusations and counter-accusations among the physicists mixed technical and non-technical claims to dismiss opponents' views – whether they felt they could 'trust' someone, whether participants had a reliable reputation within the community, and so on.

As a result of the earlier historical and philosophical studies coupled with these more recent sociological analyses of science, it became clear that scientists and their ideas should not be treated in any privileged way as being somehow free from 'social' influence. It became clear that science was in an important sense 'achieved' through social and technical negotiation, interpretation and recognition, just as other systems of knowledge are. Clearly, science may certainly strive to be the most 'objective', most rational, and thus the most reliable form of knowledge. But since there are no unequivocal rules to which scientists must conform to achieve this, the *socially constructed* nature of this most complex and interesting social institution has to be recognised. We shall return to a more detailed exploration of this position in the next chapter.

Conclusion

We have seen in this chapter how there are new patterns emerging in the character and direction of science and technology in industrialised society. These trends are only just beginning to be explored in any empirical detail by social scientists. At the same time, we have seen how the very analysis of science as a social institution has changed dramatically over the past thirty years, from the Mertonian image of an asocial, independent science, what Mulkay (1979) has called 'the standard view of science', to the significantly revised image of science as a socio-cultural phenomenon. Mulkay has provided one of the more detailed accounts of how this revision took place, illustrated by his own empirical studies on astronomy (1976) and chemistry (1985).

In short, the character of science and that of the sociology of science have both changed. Had these changes in sociology's approach to science not taken place, it is unlikely that the analysis of the emergent trends in science could be as rich or as critically constructive as it promises to be. Problematising the 'authority' of science, locating knowledge-claims in their social context, articulating the relationship between such contexts and wider economic and political processes all allow for a deeper understanding of the four trends identified earlier in the chapter.

These trends also require new policies for science, to manage its changing character. Current research in the sociology of science promises to make a positive contribution to debates in this area. Indeed, as will be seen later, recent work has already suggested ways in which this is possible, such as in the fields of health economics and environmental risk assessment. The next chapter provide a more detailed discussion of the key themes underpinning the contemporary sociological analysis of science and technology.

2 Sociology of Science and Technology

Introduction

We saw in the previous chapter that scientific knowledge can no longer be treated as some asocial, privileged account of nature. Sociologists and historians of science have shown that the general shape and direction of science is influenced by social processes. So too are the very knowledge-claims of science itself. However, while there is general agreement about this, sociologists have differed over the way this process is to be understood and the methodological implications it has. This debate provides the primary focus of this chapter.

In general, all agree that we ought to adopt a position towards science which Knorr-Cetina and Mulkay (1983) call 'epistemic relativism', meaning that 'knowledge is rooted in a particular time and culture', and that 'knowledge does not just mimic nature'. Science should not be regarded as some sort of fixed body of (objective) knowledge, closed, finished and intact – the sort of image typically presented in textbooks of science. Instead of seeing science as a 'black box', a closed, authoritative system, sociologists approach science as a *continually constructed* system of ideas, as Latour (1987) says, as a 'science in the making'. Sociologists have, in other words, sought to open the black box, to reveal the uncertainties, the negotiations, the dilemmas and controversies that inform, not exceptionally but as a matter of course, the very making of 'science'.

Since the mid-1970s, such an approach has produced a large number of empirical studies of scientific controversy, of everyday life in the laboratory, of the way in which scientific literature is constructed to present a sanitised and desocialised image of objectified knowledge, and so on. However, this work has been produced by those who adopt differing approaches in their assault on the 'black box'.

These can be summarised as:

the 'interests' approach
the ethnographic/discourse analysis approach

Even within these schools there is some debate and there are other perspectives which do not fall happily within any one of the traditions. For example, the feminist critique of science and technology has rarely entered into debate with these approaches, though it does share the overall aim of challenging the 'black box' authoritative image of science. Feminists have shown how 'scientific' conceptions of women's biology are informed by patriarchical assumptions (Rose and Hanmer, 1976; Birke, 1986), and have offered detailed critiques of those technologies most closely associated with woman as the bearer of children, initially focusing on contraception and childbirth techniques and more recently on the growth of reproductive technologies, such as *in vitro* fertilisation and ultrasound scans (Stanworth, 1987; Corea, 1985). More general feminist analyses (Cowan, 1979) link the development of a (patriarchal) science to the structural processes of capitalism. The feminist concern over science cannot be properly dealt with in the context of the approaches outlined above: it will be explored later in the book. Let us here turn our attention to the three approaches introduced above.

The interests approach

Associated with the empirical studies of a group of sociologists and historians of science based at Edinburgh University and now shared by many others, the interests approach argues quite simply that knowledge-claims made by scientists will embody or be informed by certain social, sometimes political interests. These may reflect the disciplinary, professional or more generally ideological interests and objectives of scientists. They shape ideas and are implicated in scientific debate, most clearly in scientific controversy. Studies by Shapin (1979), Barnes and MacKenzie (1979) provide some of the more detailed illustrations of the interests approach. Studies of debates over 'deviant' or 'fringe' science have made particularly valuable contributions.

Parssinen (1979), for example, has explored the historical career of 'hypnotism' within medical science. Hypnotism derived originally from Franz Mesmer's theory of animal magnetism (published in Vienna in 1776) but his ideas faced strong opposition from the Viennese medical schools and scientific academies, such that, by the turn of the century, he himself had lost all credibility and had been denounced as a charlatan. The subsequent association of the technique with popular entertainment in the 1830s and 1840s did little to improve its already battered status within medical circles. The idea of being able to control someone in a trance also offended Victorian sensibilities: as Parssinen says,

> [Mesmerism] assumed an even more sinister meaning when the operator was an older man, and the subject, as was usually the case, a young girl . . .. As mesmerism spread across the breadth of the kingdom, outraged Victorians railed against it, as though an army of seducers had been set loose to prey on the unsuspecting virgins of the land.

However, the intervention of James Braid, a medical practitioner working in Manchester, had a significant impact on the attitude towards the technique and the theory that underlay it. Braid was the first member of the established profession to claim that the trance-like condition was a product of suggestion. Rather than referring to animal magnetism, Braid called the phenomenon 'neurypnosis' and its study 'neurypnology', literally the analysis of nervous sleep. Mesmerism was thereby replaced by the language and theory of a neurophysiological concept of hypnosis. Braid's ideas received increasing support both in the UK and on the Continent where they were developed further by physicians such as Bernheim and Charcot. The eventual (1893) decision by the British Medical Association to pronounce in favour of hypnotism as a legitimate area of research reflected not so much a marked change in the theory of hypnotism *per se* but in the strength of established medicine: as Parssinen says,

What had changed was not so much the theory or even practice of hypnotism, but rather the professional community that received it. By the 1890s, the British medical profession was established legally rather than struggling for acceptance . . . . In short the acceptance of hypnotism as medically respectable in the 1890s and after is a measure of change, less of the theoretical differences between mesmerism and hypnotism than of the boundaries of the profession from the 1840s to the 1890s.

Shapin (1979) studied another nineteenth-century debate, this time over phrenology – the belief that one could determine through an examination of the cranium and the brain the existence of certain types of psychological and emotional characteristics of a person. Brain structures and tissue were linked to certain types of predisposition – towards violence, sexuality, rationality and so on. As with many other scientific disputes surrounding classification in the nineteenth century, many resisted and dismissed the phrenologists' claims, not, according to Shapin, because they had a more rational or necessarily better set of ideas but because phrenology challenged their political or occupational interests. He shows, for example, how professional medics attacked phrenology because it threatened to usurp their monopoly over diagnosis. He concludes that the debate shows very clearly the way in which technical and social interests were 'inextricably implicated' in the views of the differing protagonists: they were at work simultaneously in the judgement of competing knowledge-claims.

A species argument

Dean's (1979) study of a disciplinary dispute within botany over the proper classification of plant species is another good illustration of how the determination of what is and is not 'good science' depends on the interplay of social and technical claims made by participants in the controversy. In this case, Dean explores the dispute that has figured in the twentieth-century botany between traditional plant taxonomists who follow a Linnaean method of classifying plants into distinct species and more 'moden' plant geneticists who discriminate between plant species in terms of gene-boundaries between them.

Traditionally, the Linnaean classification is based on a plant's overall external appearance, its morphological structure, placing plants into various families or species in terms of common organic features. Most of the time, this method does in fact accord with the geneticists' approach, even though the techniques that the two botanical specialities use are very different.

However, some of the time, there *is* disagreement over where species boundaries lie, just what is and is not within the same family of plants. Dean shows how in these cases, 'nature' – the plants themselves – can support *both* classificatory schemes. That is, the evidence provided by observation of these particular types of plants, appears to lend support to both interpretations of their species membership, *even though these interpretations conflict with each other*. As Dean says, 'In this sense the natural order sustains both taxonomies; neither can be said to involve a distortion of the real facts; neither can be said to be erroneous. Nature does not in itself allow such an evaluation to be made' (p. 226).

Dean argues that this tells us that the 'discovery' of distinct plant species does not involve some simple process of reading them off from 'nature'. Instead, plant species are 'discovered' through the application of taxonomic systems which carry different assumptions, conventions and disciplinary interests. While such discrete interests prevail in botany, there will be no general consensus over the species classification of this particular type of plant.

The interests approach to scientific knowledge is conventionally referred to as 'the strong programme' in the sociology of scientific knowledge (Bloor, 1976). That is, it adopts a strongly relativistic position, arguing that scientific knowledge is itself open to sociological analysis, and that no set of knowledge-claims within scientific debates can or should be treated as in any way 'better' or necessarily more rational than any other: it is wholly agnostic as to the truth-value of any 'scientific' propositions. Those who adopt this position believe that it is possible to show how social interests are inscribed within the very construction and defence of knowledge-claims, and how social interests determine just what particular set of ideas gets adopted as 'science' and what discarded.

In other words, it seeks to understand the circumstances through which certain ideas come to be regarded by some as 'true'.

Many sociologists of science have adopted this 'strong' programme, using it to explore a wide range of debates that have featured in the construction of particular areas of cognitive fields of scientific knowledge. As Collins (1983), an advocate of this position, puts it, compared with the earlier Mertonian tradition,

> the new sociology of science has found itself in the fortunate position of being able to generate any number of fairly small, manageable, self-contained studies. The subject has, as it were, a 'granular' structure. This has helped to maintain a vigorous empirical tradition (p.86).

Collins's own many empirical studies have typically focused on what he calls the 'core set' of scientists that are involved in specific controversies. These are those who form the small expert elite group lying at the core of any particular scientific specialty, typically located in prestigious academic and research positions. It is these, believes Collins (1981), who play the principal role in determining the outcome of scientific debate, and who must therefore form the key though not sole focus on any sociological inquiry.

One of Collins' more recent texts, *Changing Order* (1985), examines a number of cases of the construction of scientific knowledge via the activity of core sets, the principal protagonists – allies and enemies alike – in scientific controversies. Collins shows how in practice a consensus over scientific knowledge-claims only emerges by way of negotiation and debate that are rather different from the supposed rigorous procedures laid down by the 'scientific method'. This is not because scientists are being in some way 'unscientific' or lax in following the (special) rules of their particular game. On the contrary, says Collins, science *can* only make perceptible 'progress' precisely because its practitioners draw on a wide range of socio-technical strategies to produce an ultimate consensus out of controversy. Most notably, Collins demonstrates that one of the principles that lies at the heart of the scientific method, the replication of previous experiments to test knowledge-claims, is in fact a methodological impossibility: experiments cannot replicate prior experiments exactly since any one experiment embodies specific skills – typically 'tacit' or hidden – which others do not exactly share: hence, 'it can never be clear whether a second experiment has been done sufficiently well to count as a check on the results of the first. Some further test is needed to test the quality of the experiment – and so forth' (p.2). This leads to the 'experimenters' regress', where each test of a prior experiment is itself susceptible to testing, and so on. Scientists endeavour to stop this (potentially infinite) regress through tactics that close off the debate: Collins's empirical studies on gravitational waves and on parapsychology illustrate his argument most clearly.

Collins himself can perhaps be regarded as a member of the core set of sociologists that have been involved in the debate over the iterests approach in the sociology of science. He and his colleagues who have been associated with him at the University of Bath (e.g. Pinch, 1981; Travis, 1986), together with the Edinburgh group, form the principal UK allies through which this approach has developed. As we shall see later in this chapter, it is also quite possible to link this body of work with other US-based socioogists of science who have been interested in examining the role of professional and disciplinary interests in determining the outcome of science policy disputes, such as Nelkin and Gillespie. The interests approach can look beyond internal debates within and between scientific specialities to debates over scientific knowledge when it presents itself as scientific 'expertise' in public settings, such as Public Inquiries. Yet core sets not only involve 'allies', as Collins himself has shown, but also 'enemies' who contest their claims. It is to such critics – fairly friendly 'enemies' in the main – that we now turn.

Critique of the 'interests approach'

Advocates of the interests approach have produced an impressive range of empirical studies exploring the social processes that inform the production of what is regarded as 'scientific knowlege'. However, they have been subject to criticism from those who argue that the 'strong' relativism of their approach is in fact *inadequately* relativist.

Critics, such as Woolgar (1981, 1988a), argues that though the interests approach is right to show how judgements about knowledge are always socially informed, its supporters fail to recognise that their *own* judgements about science are similarly mere constructions, versions of 'reality' that cannot be given any special authority. Hence, their attempts to produce *explanations* of the way certain ideas are accorded 'scientific' status, are just as much socially constructed as any set of knowledge claims. Any sociological accounts that are offered as explanations for the status of phrenology, parapsychology, plant classification systems or whatever, are ultimately based on an unwarranted assumption: the assumption that it *is* possible to produce an authoritative statement about 'the world' that does *not* embody any particular interests-position. This is, of course, the very assumption that the advocates of the interests approach want to challenge in their deconstruction of science.

Woolgar (1988b), Gilbert and Mulkay (1984) and Knorr-Cetina and Mulkay (1983), among others, argue that the sociology of science must abandon such an assumption, recognising instead that any analysis it produces is merely yet another possible *representation* of 'the world': it has no special authority. They call for a strong *reflexive* attitude towards one's claims about reality. Any socio-logical pronouncements one makes are constructions which not only can be but must be *de*constructed through a self-reflexive process. These epistemological and methodological views lie at the heart of the second major school within the sociology of science today, that is, the approach of ethnography and discourse analysis.

Ethnography and discourse analysis

The attempt to construct here a reliable account of this type of analysis is somewhat ironic. One – no, 'I' – set myself the task of producing a representation of some sociologists' views that I present as an 'authoritative' representation, as, in some way, reliable and definitive. This is ironical inasmuch as I have to undertake a task which is deemed quite improper by these same sociologists when they write about science. That is, adherents of ethnography and discourse analysis (EDA) would deny that it is ever possible to 'tell it like it is' in science – or anywhere else for that matter, *including EDA itself* – precisely because to attempt to do so requires a selective attention to detail, a blindness to alternative versions of events and the imposition of one's own interpretation of discourse and action on social behaviour. Indeed, they treat their own accounts of science in a way that is so radically reflexive that they deny that their commentaries have only one meaning (see Woolgar, 1988b). Given this, one has to accept that what follows is only my version of EDA.

The origins of EDA lie in the wider sociological approach known as ethnomethodology (see Garfinkel, 1967). Ethnomethodologists explore the techniques that people use in their everyday life to accomplish successful social interaction. Analysts focus then on the context and dynamics of social encounter and the signs and symbolic resources, especially discourse, that makes it possible. But they deny that they are thereby producing definitive accounts of this interactive drama, only their own understanding of the processes at work. Reflexively, the logic or interpretive rationale of this analysis is laid bare for others – the reader of the text, for example – to judge on the merits of the analyst's interpretive strategy, which may or may not thereby be perceived as being *itself* successfully accomplished.

This technique, developed over the past twenty years or so, has produced a vivid and lively sociological analysis as well as raising serious methodological questions over the more typical – less reflexive – strategies of sociological research. Within the sociology of science it has been developed through ethnographic research to produce finely-drawn analyses of the interaction between scientists, most especially their discursive 'repertoires', used to negotiate the status of knowledge-claims.

One of the earliest detailed ethnographies was that of Latour and Woolgar (1979) which provided an analysis of 'laboratory life' at the (US) Salk Institute in California.

There have been many subsequent case studies of 'science in the making' at the level of the laboratory (e.g. Goodfield, 1981; Zenzen and Restivo, 1982). All endeavour to demonstrate the process through which propositions, claims, or ideas of scientists come to take on – for a time at least – the status of 'facts'. A whole range of strategies and resources is required to persuade others of the 'truth' of one's claims. Sometimes these 'facts' are questioned and reconstructed in new ways as scientists deconstruct 'ready-made science'. One implication of this is that what counts as a scientific 'discovery' can no longer be regarded as in any way a straightforward recording of an obviously innovative or novel phenomenon. What is and is not discovered is itself a matter of negotiation. This process is well illustrated by Woolgar's (1976) account of the 'discovery' of 'pulsars'. (See the boxed text.)

'Gonna make you a star . . .'

On 24 February 1968, *Nature* carried an article co-authored by Hewish, Bell and three other members of the Cambridge radio astronomy group. They claimed to have discovered unusual rapidly pulsating radio sources (only subsequently referred to as pulsars).

The straightforward announcement in *Nature* conceals an extraordinary complexity in the accounts and recollections of participants. It is quickly evident, for example, by examining in detail both written and verbal accounts of the discovery, that there are clear discrepancies in participants' recollections of events leading up to the discovery . . .

Participants themselves variously reconstructed events leading up to the discovery in ways which seemed to them to produce a 'logical' sequence, but which was often quite different from other available accounts. This uncertainty about the sequence and done of the discovery provided the tangible focus for allegations [by astronomers elsewhere] of undue delay and secrecy [in making the announcement].

. . . Only in retrospect can we (do we) 'rcognise' that all along the participants were on the track of the same object: a 'pulsar' . . . Before the very possibility of an 'it' had begun to stabilise, the object (and the non-object) enjoyed at least five separate incarnations:

(1) an unusual trace; a non-object
(2) possible interference
(3) a temporary flare-up or unusual interference
(4) communication from another civilisation (little green men)
(5) new kind of pulsating radio source.

[T]he definition of the object has enjoyed a chequered career since the discovery announcement. It has been a white dwarf star, a rotating neutron star, a neutron star with a satellite, the plasmic interaction between binary neutron stars and so on. For the record, the current consensus at this level of definition favours a rotating neutron star. Should this temporarily stable construction be overthrown (revised) in the future, the usurpers will have to deconstruct nearly twenty years' worth of mobilisation of resources and argument.

Source: S. Woolgar, *Science: the Very Idea* (1988)

The temporary stability of scientific facts and the way they might be deconstructed by sociologists has been explored most fully by Latour more recently (1987). Latour argues that all science is Janus-headed: it presents itself in one (public) way as firm, solid reliable knowledge about which all agree, while simultaneously experiencing (private) uncertainty, controversy and debate. Latour unpacks and reveals the uncertain, negotiated status of the solid 'ready-made side' that confrotns us in scientific literature, laboratories and technologies, and shows how 'the construction of facts and machines is a *collective* process' involving persuasion, rhetoric and the marshalling of resources – such as citation of previous scientific papers to lend authority to one's view. Latour requires that we look at the career of claims and counterclaims in order to plot the transformation of science-in-the-making into ready-made-science. This transformation depends upon the *power* of scientists to use their experiments, writing, and that of others, to resist attack from critics, doubters or 'dissenters'. Thus, for Latour, the 'objectivity' of science derives from its rhetorical power:

'Objectivity' and 'subjectivity' are relative to trials of strength and they can shift gradually, moving from one to the other, much like the balance of power between two armies (pp. 78–9).

Through a large number of very varied illustrations from within and outside 'science', Latour shows how knowledge-claims are *constructed as facts* through their originators establishing alliances, networks of association with significant others who will lend their authority to what is being said, without at the same time trying to change what is being said: 'The paradox of fact-builders is that they have similtaneously to *increase* the number of people taking part in the action – so that the claim spreads, and to *decrease* the number of people taking part in the action – so that the claim spreads *as it is*' (p.207).

Latour's work dismantles or deconstructs the solid-state image of science and its related technologies or what he prefers to call together 'technoscience'. In doing so, he offers a number of methodological principles that the sociological sceptic needs to adopt in 'opening the black box'. Perhaps the first of these is the most important: 'We study science *in action* and not ready-made science or technology; to do so, we either arrive before the facts and machines are blackboxed or we follow the controversies that reopen them' (p.258).

In keeping with the demand for reflexivity, Latour turns his analysis of science on to sociology itself: sociology through its own technoscience of surveys, questionnaires, archives, and so on, tries to create a black box called 'society': moreover, sociology needs to muster all its own allies to establish its credentials and authority over alternative versions of 'reality': the authority of the professional sociologist relies heavily on 'textbooks [like this one]', chairs in universities, positions in the government, integration in the military, and so on, exactly as for geology, meteorology or statistics' (p.257).

Outside the ethnographic work of analysts such as Latour there has developed a parallel and methodologically similar analysis of wider technologies and the 'systems' in which they are implicated (Hughes, 1983). This work, generally known as research on the 'social construction of technology' or SCOT, challenges the conventional view that technology is self-determinant and pre-given as it unfolds over time to answer the needs of society. It questions the idea that technological development has occurred through a logical, rational self-selective path. Instead, socio-technological systems, be they expressed in the shape of nuclear power plants or domestic appliances, are always constructed and reconstructed: a technology as apparently 'obvious' and 'simple' as the bicycle has had many different possible futures – shapes, sizes and purposes – reflecting different social interests that underlay its development (Pinch and Bijker, 1987). As MacKenzie (1987) says, 'Systems are constructs and hold together only so long as the correct conditions prevail. There is always the potential for their disastrous dissociation into their component parts. Actors create and maintain systems, and if they fail to do so, the systems in question cease to exist' (p.197). It can be seen, perhaps, that those who have developed the SCOT analysis combine elements of the constructionist ethnographic approach with the 'interests' approach described earlier: successful technologies are 'constructed' through a process of strategic negotiation between different groups each pursuing its own specific interests.

Discourse analysis

Discourse analysis (DA) is a logical development from the reflexive ethnography sketched above: its focus is on the actual discourse – the speech and its texts (e.g. the published article) of scientists themselves. It is most closely associated with the work of Gilbert and Mulkay (1984) who argue that scientists' talk about their science cannot be used unproblematically as a resource by sociologists wanting to present some particular 'reading' of what is going on in science. To do so is to distort the variability in scientists' interpretive repertoires and to assume that the things they say point to the existence of 'something else' which the sociologist is convinced is 'there' – say, a particular set of professional interests. Sociologists need instead to produce (reflexive) accounts of the strategies scientists use when talking (or writing) about their work: these will be not only highly varied but often inconsistent, context-bound, and unpredictable. We cannot therefore impute any particular set of beliefs or behavioural dispositions to scientists by virtue of what they say. What then, does their discourse actually reveal? According to Gilbert and Mulkay it shows various interpretive forms or repertoires that scientists use to construct what are, for the scientists, acceptable measures of true and false knowledge. It provides 'a natural history of social accounting' which gives 'a wide-ranging description of the contextually variable methods which scientists use to construct versions of their action and belief' (Mulkay *et al.*, 1983, p. 199).

Gilbert and Mulkay argue – reminiscently of Latour – that this form of analysis should not be limited to science alone but extended and applied to *all sociology*. They argue that 'discourse analysis is a necessary prelude to, and perhaps a replacement for, the analysis of action and belief' and that this argument 'applies equally to all areas of sociological inquiry' (p.191).

Response to EDA

Not surprisingly, there have been many sociologists from within the interests model approach who have responded vigorously to the EDA challenge. Criticisms have focused on the zealous reflexivity of EDAs, implying that it moves attention too far away from scientific activity itself towards what we say about such activity. This has been argued most strongly by Halfpenny (1988; 1989) with reference to discourse analysis (DA). He suggests that those practising DA are so concerned *not* to produce a definitive account of science 'they make it impossible for themselves to tell their readers about science'; why, he asks, should DAs then write about science: why not write about what they 'say to their neighbours over the garden fence, or their shopping lists?' (1989, p.150). Others argue that, despite claims to the contrary, EDAs do in fact impose their own interpretive analyses on scientists' discourse and behaviour, do engage in selective observation and reporting: this is a chronic and inescapable feature of all and any analysis. And at the minimum, EDAs can only produce an 'understanding' of discourse precisely because they share in some of the symbolic and linguistic world of the scientists: as Collins (1983) says, commenting on Gilbert and Mulkay's DA study of a controversy associated with 'oxidative phosphorylation', they must 'understand enough about science, and about oxidative phosphorylation to know when a piece of discourse is about that subject . . . [and not] say "The Bluebells of Scotland" played on a comb and paper' (p.102). A similar point has been made by Shapin (1984). There is also a concern that the EDA position leads the sociology of science not only into a radical epistemological agnosticism but also thereby into a radical political agnosticism or even indifference: since telling it like it is no longer on the agenda according to EDA, telling it like it may be *ought* to be is simultaneously ruled out of court: for EDA, sociology cannot be a critical science.

More recently there has, however, emerged what appears to be an attempt to bring together the insights and forms of analysis of the interests model and EDA, as the debate has mellowed and new research questions have emerged – particularly political questions about science policy (see Chapter 3) and the contribution of sociology. This embryonic synthesis is explored below in the discussion of the third approach in the sociology of science – the synthetic meta-analysis perspective – wherein, in fact, one can find sociologists who were advocates of one or other of the first two positions discussed above. We shall discuss this emergent synthesis shortly. Before we do, however, we need to pause briefly and ask what sort of response has been made to both the interests and the EDA perspectives by those who have stayed with the traditional, Mertonian sociology of science that was discussed in Chapter 1. We find many such sociologists in the United States, prominent among whom is Thomas Gieryn.

The neo-Mertonian response

Probably the most eloquent and forceful of neo-Mertonians is Gieryn who has dismissed the apparent novelty and utility of the relativist interest and constructivist EDA approaches. He argues that far from being radically new in their analyses of scientists, these more recent sociologies of science are foreshadowed by the seminl work of Robert Merton: as Gieryn (1982) says,

> [M]any of the empirical findings of the relativist/constructivist programme, when stripped of polemical manifestos and trendy neologisms, could be expected from Merton's theories, and some are anticipated by his occasional steps into empirical resarch (p.280).

Gieryn provides a number of illustrations of how Merton's classic work – for example, his study of the social dimension to the 'Copernican revolution' – heralds the sort of empirical studies of recent contributions. However, what Merton did or did not say, and how this compares with current analysis, will always be arguable. For example, Gieryn believes that Merton's identification of the norm of 'organised scepticism' in science (see Chapter 1) will undoubtedly mean scientists are uncertain about their knowledge-claims and have to negotiate their status before agreeing on their value: so what's new, he asks, about the constructivists' proposal that scientists negotiate and construct scientific certainty? Constructivists might respond that the Mertonian view implies that the negotiation of knowledge proceeds according to a number of 'rational' methodological imperatives through which the technical warrant of new ideas can be objectively determined: this is not the same sense of 'negotiation' that the constructivists identify.

We will not be detained here by the claims and counterclaims of this particular debate. What is of more value is to focus on the most important of Gieryn's more general arguments raised in defence of the Mertonian tradition: that is, that the whole aim of Mertonian analysis was (and is) to understand that which is socially and cognitively *distinct* about science compared with other forms of knowledge, such as religious beliefs or art, and which has given science such pre-eminence in modern society. This, he claims, is *the* question on which we build a sociology of science, but it is one which the relativists deny any warrant to, since they seek to deconstruct science's very claim to special authority and pre-eminence over other (non-scientific) forms of knowledge. Gieryn believes that the distinctive feature of science is that nature impinges upon science, constraining and modifying what can and cannot be said about it. This is different from, say, religion, which can sustain a wide range of interpretations and constructions of deities, faiths and actions. The relativist constructivist accounts imply that 'nature' does *not* play such an adjudicating role, favouring some rather than other interpretations of its character. The neo-Mertonians, such as Gieryn, will not accept this view: he says,

> What makes science unique, in part, are institutionalised procedures which define the intersection of the natural and social worlds. The appropriate question is not *if* the natural world intrudes in scientific constructions of knowledge, but *how* it does so in science in a different way than in religion or the arts or even commonsense (p.289).

Gieryn implies then that nature impinges on science in a way that is more *powerful* than in other forms of knowledge: putting it very simplistically, for example, one might say 'a machine either works or it doesn't, one has isolated a specific gene or one hasn't' and so on. This is a powerful argument. But constructivists like Latour might simply respond that what counts as a 'working machine' or a 'gene' is always negotiable.

But *is* it? In practice, scientists do not continually negotiate the status of their ideas, taking for granted a whole range of knowledge-claims. What we need to know is how, *in practice* negotiation is ended, how a closure of debate and scientific consensus emerges. EDA analysis does not answer this question: perhaps Latour might even claim that it is an inappropriate question since negotiation is always *in principle* possible. Those who adopt the interests model approach, however, do believe that they can show how scientists' investment in certain ideas, their use of cognitive and social resources to advance such ideas and the network of relationships they have with power-brokers or decision-makers in the wider society, are all involved in the eventual closure of scientific debate (see Shapin, 1984). While this is, I believe, a valuable way of understanding the way scientific authority is created, it is not clear that it answers the central point Gieryn makes: that is, it is still uncertain how far 'nature' informs this process of closure. Perhaps one could adopt the heuristic position which allows that, as we saw with the Dean case study earlier, nature can sustain a variety of interpretations but perhaps not an infinite number. We would then need to look at the way in which the resources of scientists and their institutions are drawn on to favour particular interpretations rather than others. Chubin and Restivo (1983) believe that interpretations of nature which have been successful in the past – that is, which have gained the general assent of others – are those which result from a 'critical, reflexive, meta-inquiry'. Not all of science is necessarily like this, but to the extent that it is, we might expect it to produce more powerful beliefs about the world than say religion.

A developing synthesis?

We have seen so far that there are different perspectives in the sociology of science. As in other areas of sociology, however, where dispute between perspectives has led to the search for a common ground (such as the interminable and often misconceived debate between Marx and Weber), so here, in the sociology of science some analysts are struggling to produce a synthesis that blends the insights and methods of the perspectives.

There are a number of reasons why such a conjoining of perspectives is sought. One is that the protracted debates within the sociology of science, important though they be, have meant that the sense in which the discipline can offer a *critique* of science and technology has been muted. Chubin and Restivo (1983) call for an approach which, while drawing on the techniques of the interests/ EDA approaches to show the constructed nature of science, goes further by showing how scientific claims might be 'ideologically self-serving', potentially unethical and serving vested interests. Thus, they say, the analyst 'wields a tool of *ethical* inquiry: the ability to prescribe how data should or should not be used in the process of science policy-making' (p.63).

A similar position is adopted by Fuhrman and Oeler (1986) who argue that we must not stop at the insights of either the 'interests' or the EDA approach. The reflexivity of sociological inquiry should not mean that we stop at describing scientists' actions and discourse and our own interpretations thereof, but should also teach us to adopt a critical analysis of the place of scientific belief in society: as they say, 'the final question must be how beliefs are established' (p.300) and whose interests they serve. As Shapin (1988) has argued, scientists' interests embody forms of scientific capital – scientific skills – which they invest in different ways, whether when working routinely, writing papers or whatever. We should endeavour to exploit a 'calculative model' of the scientist as social actor – something suggested by the constructivists, such as Latour – in order to link the analysis of the world of scientists with its wider socio-economic and political context. Key aspects of this wider context are examined in the following chapter, which explores the relationship between the sociology of science and science policy. It will be suggested that the move towards a synthetic approach outlined above will make this relationship more productive than it has been in the past.

3 Sociology and Science Policy: Opening and Managing the 'Black Box'

So far, we have presented science as a form of social and cognitive activity. But it is of course a very *powerful* form of activity: while sociologists may challenge the conventional image of science as unequivocal, authoritative, objective knowledge, this does not mean that it has fooled us, and itself – like the Emperor with no clothes – all along. Science and technology are genuinely powerful knowledge-based systems reproduced by a range of powerful social and professional institutions, such as the Royal Society in England or the National Science Foundation in the US, and encapsulated in powerful technologies such as the Cruise missile that, so we are told, defends them both. Technology and science are also powerful in a less obvious way in that they are often said to be the measure of all that we should hold true and progressive. Many science-fiction writers trade off this belief in constructing their scenarios for the future.

Authors of science fiction imagine possible futures shaped by the emergence of new technological capabilities. In this, they are doing something not that far removed from the sort of exercise undertaken by those in the business of science policymaking, except that in the latter the story is usually much less exciting with heroic deeds and scientists a little thin on the ground. In their different ways, both the writer and the policymaker take for granted the immense capacity science and technology have now and in the future as agencies of social change. That they are socially constructed does not weaken them: on the contrary, their social grounding gives them greater institutional impact, whether in the economy, health care, defence, energy, or wherever.

There is today a very strongly held belief that 'we' – meaning those within the centres of decision-making and influence – can orchestrate science and its associated technologies to benefit 'us' – ostensibly meaning everyone in society. Science policy today is big business, and science policy *analysis* is fast catching up as a critical – though not necessarily always hostile – counterpart to it. This chapter explores the development of science policy by the state and the way in which the analysis of science policy by social scientists has developed in tandem with it. Rather than reviewing the whole field of science policy studies, I am more interested here in giving an idea of how the sociological insights discussed in the previous chapter can be (and have been) brought to bear on certain key assumptions of those who fashion a policy for the development and use of science and technology in society.

Characterising science policy

In many ways, science policy programmes today in most industrial states have been developed in recognition of the four trends shaping the institutional character of science and technology outlined in Chapter 1: that is, policymakers increasingly respond to and thereby help reinforce the shifts towards interdisciplinary scientific labour in 'centres of excellence', the blurring of the distinction between 'science' and 'technology', the commercialisation of knowledge, and the increasing social demand for a regulation or steering of science. Policymakers are, however, only just beginning to deal with the questions and problems that these shifts bring. Their policies too, not merely between countries, but even within one country are not always internally consistent. Perhaps one should not expect them to be so if we adopt Charles Lindblom's (1980) view that policymaking is the art of 'muddling through', rather than a carefully considered, rationally coherent process. Others, however, might feel happier to agree with Ronayne's (1984) claim that science policy is 'the intention to influence the development of science and technology in a coherent way by authoritative and informed decisions'. Whether this intent has actually been realised, it is certainly true, as we shall see, that science policy has sought to make *itself* more 'scientific' by developing more precise quantitative measures of the efficacy of government support for research.

Formal state policies for science in the West (most notably Germany) go back to the nineteenth century, though in Britain, we can only really find evidence of this around 1900 (Alter, 1987) even if one can find examples of important cases of government patronage going back much further, such as the Royal Observatory founded in 1675, once based at Greenwich but now in Cambridge. Much of the support different states gave was related to their economic and political interests in expanding empires in the latter part of the nineteenth century. Hence governments of the day were more likely to foster geographical and geological research, in part related to resource interests in the newly-acquired colonies. There were, of course, other more parochial but just as important concerns within nations to do with health care and food production that led to the establishment of centres of scientific activity in public health laboratories and agricultural research stations.

But the development of a more extensive policy of state support for science followed the period immediately after the First World War. In Britain, there was considerable unease about the relative technological superiority of both Germany and America. As the government declared – in rather tortured, civil service style – 'There is a strong consensus of opinion among persons engaged both in science and in industry that a special need exists at the present time for new machinery and for additional State assistance in order to promote and organise scientific research with a view especially to its application to trade and industry' (HMSO, 1914–16, p.351).

Despite its rather dry formulation, this link between science, technology and the economy is the principal *raison d'être* for a formal science policy. The state also has a major interest in science policies for military and social purposes (e.g. education and health) though these vary considerably from one country to another. All states have sought to develop rigorous criteria for assessing the relationship between what they put in and what return they receive on behalf of their own 'national interest'. However, there are difficulties in deciding what is socially optimal, especially in the more politically contentious areas like weapons expenditure or distribution of resources within health care systems. Moreover, as Clark (1985) argues, managing science and technology to serve the national interest is made more difficult by the 'technological and commercial uncertainties' of innovative developments in new fields, such as superconductivity, information technology or bio-technology. Freeman (1974) would argue that uncertainty is an inescapable feature of industrialisation at any time and not just confined to the high-tech sectors of an economy. Rothwell (1984) has also discussed the problem of determining the efficacy of any one country's national innovation policy, noting that 'the greatest problem of innovation policy is that it has been more an object of faith rather than of understanding' (p. 164).

The question of how innovation is itself encouraged, measured and evaluated is a crucial dimension of science policy analysis today. It is one which has been of growing interest over the past two decades to social scientists, especially economists, interested in understanding the factors influencing technological change, and more specially the transfer of technology from the research (R) bench, downstream to the 'development' (D) bench (see, for example, Coombs *et al.* 1987). An increasing number of senior managers of small and large companies have found social scientists knocking on their doors – the scientists having negotiated the security gate – to find out about their in-house strategies for managing R&D.

The central concern of industrial states has been to determine the best way to 'exploit' their national knowledge bases. In Britain, for example, the mid-80s saw a growing trend towards emphasising the role of 'strategic' R&D in science, that is, research which is applied but 'in a subject area which has not yet advanced to the stage where eventual applications can be clearly specified' (Cabinet Office, 1986). The promise of 'strategic' research in certain areas was given detailed consideration by two science-policy analysts, Irvine and Martin (1984), in their text *Foresight in Science*. The exploitability of strategic research was highlighted in what was then the UK government's Advisory Council for Applied Research and Development (ACARD) report *Exploitable Areas of Science* (1986) which recommended that 'a process should be established for identifying exploitable areas of science, which has some certainty of continuity, for the long-term economic health of the country' (p.12). Subsequently, a part-state, part-industry funded organisation was established (at a cost of £5 million), the Centre for Exploitation of Science and Technology (CEST) initially based at the Manchester Science Park, which at the end of 1989 produced its first advisory report on what might seem to be the rather unglamorous subject of adhesives.

The notion that science and technology could be externally directed by carefully-crafted state policies was given one of its most interesting airings in West Germany during the mid-1970s in the so-called 'finalisation debate'. Briefly, a number of German historians and philosophers believed that certain areas of science were theoretically mature, 'finalised' so to speak, in the sense that they had answered most of their theoretical and empirical problems and were as such susceptible to external direction. As Bohme (1983) puts it: when a field has achieved theoretical maturity, 'research fronts can be planned . . . with certain goals in mind. Further theoretical development within the discipline can then proceed broadly along the path indicated by such external goals. We term such a process "finalisation"' (p.9). Typically, these fields, such as agricultural chemistry and fluid mechanics, had been most likely to be exploited by private capital for profit rather than by state agencies for public welfare. The advocates of this thesis argued that mature fields should be directed primarily for the public good. Perhaps it was not surprising, then, that finalisation was branded as a neo-Marxist threat to the autonomy of science (see Schafer, 1983).

More common today in capitalist industrial states is the desire by government to encourage an economic environment which will be most conducive to the *transfer of technology* from centres of research in academia – universities, polytechnics – or government research establishments to the development labs of the private sector. The phrase 'technology transfer' became the warcry of academic and industrial liaison officers throughout the length and breadth of Europe from the early 1980s. Many within Europe believed that the Japanese had solved the problem of technology transfer so effectively that their new industries would be able to take advantage of the Europeans' greater flair in transferring their capital abroad, buying sophisticated Japanese imports.

Assumptions of science policy

Clearly, science policymakers in government, and their advisers elsewhere, will endeavour to shape their policies for science in such a way as to optimise the best conditions for its exploitation. This might be called the rational model of policymaking: choosing the most appropriate means of achieving the goal of exploiting the science base to the full. As we saw earlier, however, policymaking is often difficult to determine and perhaps even more difficult to implement: if one were to examine in detail some of the twists and turns of policy in, say, the UK it would be difficult to adopt a conspiracy theory of history. There are many contingent factors, political, economic and, more broadly social, which can detail the most apparently 'rational' plans. Yet, of course, policymakers must at least publicly justify their decisions in terms of some ostensibly rational criteria.

Typically, to do this, they might be tempted to treat science and technology as 'black boxes' that can be managed on behalf of the national interest. The very extensive literature, for example, on how best to transfer technology from the research lab to the market, or more widely from First World to Third World countries for 'development' purposes, tends to presume that *what the technology actually is* – whether techniques, ideas, artefacts, or knowhow – requires limited consideration. One is, as it were, simply transferring the 'black box' from one context to another. This is especially true of contributions from those working in the industrial liaison field between academia and the corporate sector for whom most of the time the problem is one of determining the best mechanisms for transferring technology from the educational establishment into industry, a linear view of the relationship between scientific invention and technological application.

Or again, science policymakers rely quite properly on obtaining the best possible advice they can get prior to taking a decision on a perhaps controversial issue. Public planning inquiries in Britain on nuclear power involve the careful and often extremely prolonged presentation of evidence from different expert witnesses: the Sizewell B Inquiry is a recent notable example. Again, the policymaker has to presume that the expert witnesses' contributions carry the authority and integrity of the scientific fields which they represent. As inquiries reach their conclusion there is a tendency to 'black box' expertise itself to justify the recommendations of the inquiry. This is perfectly understandable, and it is something we shall explore more fully in Chapter 6. But the notion of the *expert-as-authority* is clearly another important assumption of science policymaking.

A third feature of science policy is that there be some way in which one can assess the quality and quantity of the research output of a particular field which the state in various ways is trying to support. This tests the wisdom of earlier decisions and assesses the merits of plans for the future. Again, this is a most rational requirement of any planning for science that can claim some sort of monitoring and self-regulation, particularly in areas where resources are limited. But again, there has to be the assumption that *one can produce objective criteria – quantitative and qualitative – to allow objective evaluations to be made.* There have indeed been very careful and sophisticated measures that have been developed as 'performance indicators' of scientific fields. Both government agencies, such as the US National Science Foundation's *Science Indicators* (published since 1973), and academic science policy researchers (such as Irvine and Martin [1984]) have developed an impressive range of techniques that perform two tasks: the *collection of information* databases recording the productivity and pattern of scientific research and the *evaluation of research* through a number of output indicators. Some of the work at the Science Policy Research Unit (SPRU) at the University of Sussex has been concerned with an evaluation of 'Big Science' programmes in a number of countries, that is programmes that support basic research in such areas as astronomy, space research and high-energy physics (particularly nuclear particle accelerators). The SPRU researchers have asked whether it is right to devote considerable resources to these areas at a time when science budgets are generally static. One cannot rely on the judgement of senior scientists – the system of 'peer-review' – since the extent of these programmes is likely to mean that most of the relevant scientists will be involved anyway, hardly a recipe for independent assessment. Other criteria, therefore, have to be developed to determine whether these 'big science' fields are being exploited effectively and productively by those that work within them: various criteria have been suggested including the number of scientific publications produced by those in the field, how often these are cited by scientists elsewhere, how others rank workers in the area and so on.

Performance indicators: what they tell us

Commenting on one of their studies of the value of research output produced by those working at the British Isaac Newton Telescope (INT), Martin and Irvine note that . . . if a telescope like the INT produces relatively few papers at high cost which receive comparatively few citations in total (even though each paper on average has a reasonable citation-per-paper figure), and if it yields rather few highly-cited papers and is ranked towards the bottom of a list of 12 telescopes by 50 astronomers, we would be reasonably confident that its scientific performance had not been particularly good in world terms.'

from B. Martin and J. Irvine, 'Evaluating the Evaluators: A Reply to our Critics', *Social Studies of Science*, Vol. 15, 1985 (pp.568–9)

One of the most important points which Irvine and Martin have consistently made in their work has been the importance of using a combination of indicators, because of the shortcomings of each one taken separately. Thus, in relation to the assessment of the comparative performance of research institutes working in the same scientific field, they have used a combination of measures including scientific papers, citations, most-cited papers, and the peer judgement and rankings of fairly large numbers of younger scientists, as well as the traditional method of peer judgement of senior scientists. Where these indicators all converge, then this can provide a rather stronger foundation for decision-making than the simple use of peer judgement by committees of senior scientists and officials. Where the indicators do *not* converge, then they can raise important questions for discussion, such as why the papers of a particular institution are hardly ever cited in the international literature, or why younger scientists have a rather different assessment of the quality of a particular institution than the more senior people.

Source: C. Freeman, 'Quantitative and Qualitative Factors in National Science and Technology Policies' in Annerstadt and Jamison (1988).

The final assumption of the science policymaker, regarded perhaps as a necessary adjunct to the promotion of science, is that its growth can and must be made to be accountable to 'the public'. Regulatory regimes and forms of accountability vary from one area of science and technology to another and from one country to another. What may be scrutinised and banned in one country may be entirely ignored by state agencies in another. Precisely because of this, the pressure for regulation often comes from political or public pressure groups concerned about the impact of specific technological developments in their country. As we shall see in the next chapter, the state's response to such concern is also variable, reflecting distinct political cultures within which the regulatory process has to operate. In the United States, for example, there is a much more open, contested and gladiatorial style of policymaking, to which decision-makers react by 'playing safe', typically going for strong regulation where there is doubt over the effect of a technology. In Britain, on the contrary, the discrete and less overtly politicised nature of control of science allows a much looser regulatory rein over its development. As we shall see in Chapter 6, these political differences are very clearly illustrated in the two countries' approach to the development of genetic engineering and its attendant risks. Notwithstanding these differences, it is generally assumed that science and technology should be made to operate in some sort of regulatory framework.

We can see, then, that there are four principal assumptions underlying the rational model of science policy: that scientific knowledge can be treated as a 'black box' and managed on behalf of society; that expert authorities can advise on this process; that indicators of how successful this process is are available; and finally, that adequate regulatory safeguards can be built into the whole process.

Problematising science policy assumptions

While one can understand and indeed applaud these basic tenets of science policy, sociologists of science have shown through numerous empirical studies over the past decade or so that these assumptions can only be sustained by glossing over what is in fact a very messy, uncertain business. It is not surprising, though, that this 'glossing' takes place: science policy is primarily a prescriptive and practically-oriented task that needs, perhaps, to make certain gross assumptions in order to get on with the decision-making role it performs. In Britain, the two main policy-advising agencies, the ABRC (Advisory Board for the Research Councils) and ACOST (the Advisory Committee on Science and Technology), produce regular reports and recommendations to government on the basis of what they have to regard as sound – though undoubtedly never complete – information and advice from senior members of the corporate and academic research communities. It is therefore not surprising that members of these – and similar – committees might hesitate to question the assumptions on which they seek advice and on which they give advice. Their job would be made much more difficult, some might believe, impossible.

It is increasingly common, however, to find sociologists of science arguing that policymaking might in fact be strengthened were it to pay closer attention to their research findings. Even if one were to accept Martin and Irvine's (1985) warning that 'it cannot be assumed that what is most interesting from a sociological point of view is necessarily most relevant from a policy perspective' (p.572), there is still much that a sociological analysis can offer: most importantly, it can encourage those involved in policymaking and evaluation to be more reflexive about the assumptions they make. Sociology should help to refine the instrument of policy without becoming a slave to policymaking: it must be capable of offering a critique of policy, to limit the extent to which misplaced assumptions produce ineffective policies. As Cozzens (1986) has argued, 'The insights which social studies have to offer to the policy community come from non-policy research. A proper balance between policy-oriented and other kinds of studies of the scientific community must be maintained if science policy and social studies of science are to engage in a meaningful dialogue over the long term' (p.16). I can illustrate how this might be possible by looking at a few of the assumptions of policymaking outlined above and subjecting them to a brief sociological critique.

i) Problematising 'technology transfer'

'Technology transfer' means the transferance of a technology or technique from one context to another. As such, the idea has been around for a few million years. But the phrase has in recent years been most closely associated with two types of transfer: the transfer of technology from the First to the Third World, and the transfer of technology from the context of the research laboratory to that of the factory manufacturing goods and services for the market. The first has been linked with debates about appropriate technologies for indigenous development, the growth of a 'new international division of labour', the impact of ecologically damaging systems of production in rural sectors and so on, while the second has figured in the analysis of the process of technological innovation in industrialised states. It might also be reasonably argued that the first was in response to the chronic economic crisis within Third World states – first acknowledged in the mid-1960s, while the second has been associated with rather different, more recent, but nevertheless, critical, problems of the development of First World countries: namely, how to remain innovative and how to stay ahead of one's competitors (in the European and American context this means Japan). The focus here will be on this second sense of technology transfer (for a detailed discussion of the first see Yearley, 1987).

As suggested earlier, there is among those concerned with transferring innovation from the lab to the marketplace a tendency to 'black box' technology. This allows one to construct a technical boundary around a technology and thereby to envisage its technological trajectory from laboratory to market, from invention to innovation. There is, as it were, an Aladdin's Cave of precious technologies within academia that might be used to make a country rich: this is the sort of image conveyed by a British Minister in 1986 who declared that 'transfer from academia to industry involves unlocking the vast stores of research and development and embryonic technology which exist in universities, polytechnics and other higher educational institutions' (Butcher, 1987). Government may initiate various programmes for promoting and transferring new technologies to the industrial infrastructure. In Britain, such programmes (for example, the DTI's 'Link' scheme) are common features of the university–industrial liaison landscape. Academic establishments have established technology transfer units, often linking with the government's principal agency for commercialising inventions, the British Technology Group (BTG). Similar efforts have been made in many other countries and increasingly so across countries through European or transatlantic programmes (Barnes, 1988). Conferences and seminars on technology transfer fill the filofax of the industrial liaison officer. Many people regard these programmes as successful, and prizes (such as the BTG's EPIC Award) are given to those who demonstrate a flair for innovation. But is this because the technologies that are transferred have had a trajectory like the arrow that hits its target? Unfortunately, matters are more complicated than this.

A few dates for your diary . . .

26–28 March 1990, London UK

European Conference on the Profitable Exploitation of Technology. Sponsored by Defence Technology Enterprises Ltd in association with BASE International, Arthur Young, PA Technology Group, British Technology Group, Barclays Bank, Unicorm Seminars, Butterworth Scientific Limited, The Patent Office.

22–24 May 1990, Glasgow, UK

TRANSTECH EUROPE – The Second European Technology Transfer and Innovation Opportunities Exhibition and Conference.

June 1990, Vienna, Austria

Worldwide Technology Transfer Conference. Contact Technology Transfer Conferences Inc., Nashville . . .

First, sociologists have shown that at the level of the research laboratory, the transfer of a technology from one research scientist or group to another can be very difficult. This has been demonstrated most fully by Collins (1982) in his study of the attempted reconstruction and replication of Transversely Excited Atmospheric (TEA) laser by British scientists following its invention by a group of Canadian scientists in 1970. Collins found that the British physicists experienced considerable difficulties in trying to reproduce the laser, and he himself was to experience this first hand working alongside one of his respondents, also trying to build a version of the laser. Collins' study shows how important *tacit knowledge* is in scientific work: that is, the techniques and technologies of scientific research are only made to work because of researchers' experience in trying to make them work. That is, they develop skills and understanding which are hidden or 'tacit' and which, without direct contact, are difficult to transfer to other scientists working elsewhere. Such knowledge is *not* the same as that which is formally inscribed in a published scientific paper: simply having access to papers and even private laboratory notes does not make the task of replicating science – in this case, building other lasers – any less difficult. Collins demonstrates, therefore, that the transfer of knowledge is highly problematic within research field networks, arguing that 'transfer of skill-like knowledge is capricious', that it 'travels best (or only) through accomplished practitioners' and that like any skill (such as riding a bicycle) 'it cannot be fully explicated or absolutely established' (Collins, 1985, p.73).

Transferring technological 'certainty'?

The advent of DNA fingerprinting in forensic science which is used to test samples of blood or other body fluids in criminal cases to determine *without a shadow of doubt* where responsibility lies – just who was the murderer, the rapist, the real father and so on – has been heralded as an example of 'hard' science being successfully transferred to the wider society, for a small charge. But even here, we find that certainty has to be continually recreated so that sometimes the transfer is much less straightforward than might be presumed. Turney (1989) tells us why:

> Certainty is a rare commodity, but today you can buy a small portion for £122 plus VAT. This is the charge made by Cellmark Diagnostics for a single DNA fingerprinting test, with its uniquely sensitive result . . . [However], translating the certainties of the molecular biology lab into a routine test or a judicially acceptable package of evidence is no easy matter. It rests on a series of technical and legal conditions which may vary from case to case or from country to country.
>
> In this area, technology transfer does not just mean getting a promising discovery out of the laboratory into the commercial world. It also means bridging gaps between the habits of mind and standards of proof of hitherto

largely separate professional and intellectual communities
[of science and the law].

Source: J. Turney, 'Gene Believers', *The Times Higher Education
Supplement*, 3 November 1989, p.17.

The developmental 'trajectory' of any technology is, then,
problematic from the start. It is further complicated when new
social actors outside the research lab – perhaps government civil
servants or industrial investors – join the scientists in shaping the
new technology. Now commercial and bureaucratic interests in
and perceptions of the emergent technology begin to play a role.
One important early consideration they will bring to bear on the
academic 'invention' is whether it is a genuinely inventive step, and
so a technology that can be protected as novel, perhaps although by
no means only by patenting it through the Patent Office. This
determination of novelty is by no means a simple matter, however,
and sometimes dispute arises between the different parties involved
– the inventor, the patent office, companies interested in or affected
by the patent claim – over the originality and patentability of the
technology. In general, the 'success' of the technology, and what
eventually is transferred depends on factors such as the level of
resourcing for projects, how novel emergent technologies are
perceived, the degree of competition and collaboration between
commercial companies that have an interest in the area, whether
strategic decisions for the development of the technology are based
on short- or long-term considerations and so on. This is well
illustrated in Britain with the government's Alvey programme
(1983–8) designed to foster R&D in new areas of information
technology (IT), (notably Artificial Intelligence including Intelli-
gent Knowledge-Based Systems). This programme brought
together academic, industrial and government (including military)
department interests in IT primarily in the so-called 'precompetiti-
ve' – that is strategic rather than applied – stage of research. While
there have been many 'spin-off' technologies that are now commercialised, commercial interests made some of the pre-
competitive collaboration less innovative than it might have been.
For example, research projects funded by the Alvey programme
had to be conducted according to the contractual terms of the

Alvey Framework which unwittingly encouraged risk-sharing projects between companies involved. As Guy et al. (1987), put it, if two companies, A and B, 'enter a project where each pursues a different approach, and if A's results are exploitable and B's are not, then under the strict terms of the Framework B has no entitlement to A's results but A can use B's results in so far as they are necessary to exploit A's own. The resulting asymmetry is potentially a problem as participants are encouraged to fight to gain the least risky line of research' (p.108). It is difficult to say how far these commercial interests have shaped the development of technologies at the research bench because of lack of evidence, but Guy's remark suggests that they have done.

Finally, within the commercial context itself, one can find an increasing interest within companies to produce a 'technological audit' of their areas of expertise in order to determine medium- and long-range investment plans. To do this, however, one needs to develop some system whereby one can classify technologies into different types: this could be done by product area, by disciplinary basis, by production function, and so on. What one eventually chooses will have an important effect on the velocity and character of technology that is transferred within the different operating Divisions of the company. But in having to make this choice, companies are forced to unpack the technological black box, and so again, here as elsewhere, there is likely to be considerable debate about the sort of criteria to be used to determine the boundaries of technologies, their relatedness and the most appropriate mechanisms for linking them together.

From research lab through to the company, therefore, we can see that the boundaries of any technology are not clearly defined, and that its transfer to another area is a matter for continued negotiation and debate between more or less powerful interest groups. Since there is no clear trajectory to technological development, the process of technology transfer is less like an arrow and more, perhaps, like a pinball. Contrary to the linear model of technology transfer, innovation occurs through a much more complex interaction during the R&D process in terms of feedback on knowledge (both tacit and formal), technical skills, expertise and so on. Innovation in companies therefore, involves combining all these different science and technology inputs from both internal and external sources at all times during the development process.

ii) Problematising expertise

There are in all industrialised countries – and in many Third World countries too – institutions of state and government responsible for deciding policies that will promote the growth of science and technology as major contributors to economic development. As we shall see in the next chapter, the style, impact and resources devoted to these policies vary considerably and have varied within countries over time: China, for example, has recently shifted away from its policy of a self-reliant technology to one geared towards the acquisition of key technologies from other countries while encouraging as many of its own research institutes (4760) as possible to develop stronger links with state ministries and industry (Conroy, 1988).

Wherever state policy is being made – in Washington, London, Bonn or Beijing – it relies on the advice it receives from scientists working in academic, civil service or commercial sectors. Sometimes this advice is commissioned, sometimes freely given, sometimes acted upon, often shelved and even occasionally suppressed. *Ad hoc* committees of experts may be established to advise on specific matters as they arise while standing committees or more formal bodies, such as the US Office of Technology Assessment, will deal on a regular basis with strategic long-term issues, such as policy towards information technology, as well as the routine regulatory issues, such as the safety of new drugs, pesticides, aircraft design, and so on. In Britain, experts are often witnesses before Public Inquiries, required to attend meetings of the Parliamentary Select Committee on Science and Technology or asked to assist major Committees of Inquiry, such as the Warnock Report (1985) on *in vitro* fertilisation and embryo research.

The principal role of the scientist here is to give advice to policymakers, advice which will, if accepted, be regarded as authoritative and so lend weight to the decisions of government. Government is keen to give the impression that its decisions are rational, based on the best possible advice available. Clearly, this presupposes that such advice is free from prejudice or self-interest, an image of science which, you will recall, resurrects the Mertonian picture of science as 'disinterested' and 'objective'. As Jasanoff (1987) notes, in the US, 'regulators must eventually present the public with a scientific rationale for actions dealing with technological hazards, marshalling the supporting data and rejecting contrary evidence as persuasively as possible' (p.197). It is clearly important that the public be reassured that policies governing research on, say, the human genome or change in global climate are sensible, effective and ethical. The assumption is then that expertise provides a sound formation for proper policymaking.

Again, however, we find that sociological research has found that expertise and its use is what I have called *socially contingent*, that is, perceived, evaluated, and rewarded according to the audience and the context in which it appears. Thus, the actual force and role of an 'expert opinion' can be quite variable, and may have a very loose connection with policymaking. Why might this be so?

First, it has been demonstrated through the many case studies that have explored scientific controversy that what counts as 'good (expert) evidence' or advice depends on the relative *power* of different groups to define some knowledge claims rather than others as more objective and so more acceptable. The collection of studies by Nelkin (1979) demonstrates this point particularly well. Moreover, because of this, different experts will ally themselves with different interest groups whose own collective ethic is likely to influence the sort of claims supportive experts will make: this is particularly true during periods of controversial debate. As Jasanoff (1987) says: 'In areas of high uncertainty, political interest frequently shapes the presentation of scientific facts and hypotheses to fit different models of "reality" ' (p.195).

Nelkin outlines the sort of controversial areas typically associated with gladiatorial dispute between scientific experts. They include disputes concerning planning: should this nuclear power station be built?; debates over hazardous technologies: how should we assess the risk of this or that chemical/pollutant, etc?; questions over government policy; whether to deliver a form of technology or not, for example, fluoride in the public water supply; and whether, more generally, developments in science and technology compromise or more strongly threaten the values held in society at large: various areas of biotechnological research, especially genetic engineering, have already been subject to this type of critique.

Secondly, sociologists have shown how expert scientists allow their *anticipation* of what is likely to be more widely acceptable to influence the sort of technical judgements they offer those who have commissioned their advice. For example, in Britain, the

Warnock Committee, whose brief was to provide recommendations to Parliament about the scientific and ethical aspects of human fertilisation and embryo research, dealt with an issue about which there was no consensus. With regard to embryo research, the Committee eventually recommended that research on human embryos should be permissible up to fourteen days. The ostensible scientific reason given to justify this was that the first signs of a nervous system – the so-called 'primitive streak' – do not develop until then and so (presumably) no pain could be felt by the embryo. However, members of the committee were also aware of two opposing views, both represented in and outside parliament, with which their advice had to reckon. On the one hand, there were those who argue that the sanctity of human life begins at fertilisatio, and to whom therefore any embryo research is an outrage. On the other hand were those who believe that it is only at birth that one becomes a human being and for whom, therefore, embryo research for a much longer period of gestation is acceptable. The Committee made a *political judgement* that the 14-day period would be an appropriate compromise between the two groups, which had some technical warrant for it, and which was more likely than other options they considered to win the approval of Parliament and be translated eventually into law.

Thirdly, expertise has more generally been associated with broad ideological positions which favour certain interests over and against others. In the past this has been a familiar accusation levelled against those who have promoted scientific measures of intelligence and IQ (see Kamin, 1974; Rose, 1976). More recently it has been an accusation levelled at expert advisers working ostensibly as independent scientists within government departments or state agencies: this is particularly so of the controversies that have surrounded nuclear power and radioactive waste disposal. It is not simply that technical uncertainties in this area enable government to manipulate the evidence (Tierney, 1979); conspiracies are probably very rare and difficult to manage. More usually, expert recommendations for a particular technology policy embody a host of assumptions about the best way of doing things: the entrenchment of professional and institutional practices among established scientists has been well-documented and is considered in more detail in Chapter 6. The government can, of course nudge its scientific advisers in the right direction, illustrated by the following statement (HMSO, 1979) made by the British government about its policy on disposal of radioactive waste in the sea (in an area just north of the Channel Islands):

> We recommend that the UK should continue to seek to develop realistic international standards for disposal of low and intermediate level waste at sea. We believe that there can be quantitative justification for an increased sea dumping programme and we recommend urgent research to build up a body of knowledge which will demonstrate this (p.118).

Finally, recent work by Collingridge and Reeve (1986), challenges the whole notion that scientific expertise can and should feed into policymaking. Acknowledging previous research that shows how far scientific expertise can be compromised by political factors – awarding grants, for example, to those scientists likely to produce results which favour a particular position – Collingridge and Reeve argue that at the end of the day policy is made less by reference to technical advice than by a fudged compromise between the different parties involved: there will, they claim, never be a consensus among the competing scientific experts such that the ultimate political compromise cannot be said to rely on technical information. So why bother with experts in the first place?

Paradoxically, the extensive research given over to policy-related matters has one main function: competitor scientists pull the rug from underneath their own feet such that no one group will be powerful enough to shape science policy. As Collingridge and Reeve put it:

> The role of scientific research and analysis is therefore not the heroic one of providing truths by which policy may be guided, but the ironic one of preventing policy being formulated around some rival technical conclusions. Research on one hypothesis ought to cancel out research on others, enabling policy to be made which is insensitive to all scientific conjectures (p.151).

This is a surprising claim, and one which is not meant to be cynical. Indeed they would argue *against* the idea that scientific expertise should feed into policymaking as those who adopt a rational decision-making perspective might presume. Quite bluntly, they say, 'The principal thesis of this book . . . is . . . that no choices of policy are ever made which are sensitive to any scientific conjecture, and that no such choice ought to be sensitive to any scientific hypothesis' (p.28). This is not a position with which everyone agrees. Barnes (1987) for example, although accepting much of what they say about the way differing experts' opinions can often run head-on to each other in the policy arena, says this does not *necessarily* mean that we have to deny that some scientific advice may in principle be more sensible than some other. Inductively, he suggests, the view that smoking causes cancer is more acceptable than the proposition that the correlation between the two hides what might be another 'real' cause of lung cancer.

Whatever position one adopts here depends more on one's philosophical approach towards knowledge. We can, however, accept the general claim that policy decisions *in practice* are often perhaps always determined by non-technical considerations. Nelkin (1979) recognised this over a decade ago when she commented on a number of science policy controversies: there is not much evidence, she says, 'that technical arguments change anyone's mind. In the disputes over fetal research and even in the various [nuclear station] siting controversies no amount of data could resolve value differences. Each side used technical information mainly to legitimate a position based on existing priorities. *Ultimately, dramatic events or significant political changes had more effect than expertise'* (emphasis added; p.20). A similar argument could be made of the British government's nuclear power programme, specifically its decision in late 1989 to postpone plans to construct further Pressurised Water Reactors. Though this followed in the wake of the Sizewell B Inquiry (see Chapter 6), the Inquiry chairman had given his (somewhat cautious) approval to the PWR at Sizewell and so, most thought, the rest of the PWR construction programme elsewhere. However, subsequent plans to privatise the electricity supply industry – including nuclear power – have killed off any further expansion at this time precisely because commercial investors found the nuclear part of the portfolio highly unattractive.

We can see from this discussion that scientific expertise certainly has a role to play in policymaking, but it is one that will be limited by wider socio-political factors. It will, no doubt, appear to be ritualistic, as Wynne (1982) has suggested in his analysis of the

Windscale Inquiry in Britain: in fact he goes further and suggests that the purpose of science–policy inquiries in general may be to act as 'social rituals, distracting everyone from the uncomfortable recognition of moral and social' uncertainties and risks (Wynne, 1988). As Nelkin (1979) has also observed, science–policy debates can often hide political value-judgements behind technical advice especially in the unrelenting search for planning 'efficiency'.

Sociologists have shown then how expertise in a public forum has to be socially constructed where it lacks strong political or economic allies. Sociologically, therefore, it is important that policymakers establish ways in which the rhetorical use of and politically unequal access to expertise are minimised in order to prevent any one group from being able to set the research and development agenda. This raises the question of mechanisms for maximising accountability to the public. One such mechanism is the range of regulations that are instituted by government to control science and technological developments. This brings us to another of the assumptions of the policymaker whch has been subject to sociological scrutiny.

iii) Problematising regulation

Rather than review the various issues that have been explored by sociologists in connection with risk and regulation (see Brickman *et al.*, 1986; Vogel, 1986), I want to focus on one particular case study to illustrate the sort of questions sociologists might pose about regulatory frameworks surrounding science and technology. In general terms, there has been a rapid expansion of regulatory frameworks governing the development and impact of technological innovation, especially since the 1960s, as the state increasingly recognised the need to control actual and potential hazards associated with such innovation. Sometimes new regulations have emerged in response to hazardous events, such as the European Directive regulating potentially dangerous chemical installations that followed in the wake of the 1976 Seveso disaster in Italy (see Ives, 1986), or existing regulations may be reassessed as happened to some extent after the Chernobyl nuclear plant exploded in April 1986 (Hawkes, 1986). The new discipline of 'risk assessment' has grown rapidly as policymakers and academics have sought to develop a more sophisticated approach to the control and regulation of hazardous science. The growth in environmental risk 'impact assessment' research has been particularly strong since the mid-1980s, especially in the United States, where environmental regulation has been more extensive.

However, as Irwin and Vergragt (1989) have argued, much of this analysis of the impact of regulation on technological systems has tended to assume that the pattern and process of regulation is *external* to industrial innovation. That is, it is typically the case that the analysis charts the impact of regulation on innovation in a uni-directional way, looking for ways in which regulation constrains technological development and so reduces innovatory output. While there is a simple sense in which legal regulations are external to industry inasmuch as they are normally drawn up and given legislative power by government, this fails to recognise the more complex *interactive* process between a numer of different social actors – government legislature, bureaucratic agencies, industrialists, scientists, pressure groups, and so on. It is this more complex perception of regulation that Irwin and Vergragt argue we need to adopt.

They illustrate their argument through an examination of the impact of UK regulations introduced during the 1980s to control hazard in the petrochemical industry, the Control of Industrial Major Accident Hazards (CIMAH). These require plants to make emergency provision for neighbouring areas in case of major accidents, to provide extensive information on the plant, and to prepare a complete safety audit of it. A simple model of regulatory impact – what they call an 'input–output' approach – would look for the economic implications of the regulation for innovatory activity and technological development in the plant. But this assumes that *what the regulation is, how it is understood, and what effect it has* are all likely to be similar as one moves from one plant to another. The 'input', the regulation, is regarded as standard across the industry.

Against this view, Irwin and Vergragt want to offer a 'richer perspective' and suggest a more complex *interactive* model. They argue that it is important to understand the organisational processes that interpret, refine, and reconstruct the regulations. They can claim therefore that we have to consider a number of rather different questions from those suggested by the input–output approach, including:

At what levels within particular companies are
ponses decided?

To what extent within given companies have
environmental and safety protection become an int
the innovation process alongside more acknowledg. ...ues
such as profitability and growth?

Which key groups of actors have had the greatest influence over
such decisions?

What consequences has regulatory change brought about for
R&D management?

To what extent has the timing and form of regulation been
influenced by the demands of the innovation process?

(Irwin and Vergragt, p.67)

They note that companies within the petrochemical sector are not
then responding to CIMAH by merely trying to meet its requirements but are interpreting what its 'requirement' actually means for
them, and even 'rethinking chemical processes in order to anticipate future changes and adapt to the new national and international safety climate' (ibid.).

What does this say to the policymaker? Principally, to abandon
the notion of regulation as a 'reactive' strategy towards hazardous
technological systems, and to adopt instead a more sophisticated
view of regulation as a socially negotiated process which is shaped
just as much by those to whom the regulation is supposed to apply
as it is by those who formally draft it. Were one to analyse this
process more fully, policymakers might be in a better position to
produce regulatory frameworks that were sensitive to the local
context and so act as 'positive' means of technical control 'rather
than [being viewed] simply [as] restricting the operation of the
market-place' (p.58).

Summary and conclusion

The emergence of a complex of policies that both promote and
steer science and technology is a result of the changing industrial
and political environment of the past decade. On the economic
front, First World industrial capitalism has undertaken a massive
restructuring of its activities, moving towards an information-

based, flexible system of national and international production, though the speed and extent to which this is happening is much contested. The social costs of extensive unemployment and poverty associated with these changes have often been hidden behind the rhetoric of market economics. The new technologies which have been associated with this – information science (especially software), biotechnology and new materials science – have been regarded as the areas most likely to stimulate a period of renewed growth for industry. The growing flexibility of companies means that they need to target their science and technology research to specific national and international markets.

Multinational companies' global strategies for R&D have been matched by the growth of national and international structures that both enable this restructuring to take place, but also review and to an increasing extent regulate it. Policies for science, within and between countries – such as those operating across the member states of the EC–are only part of this process, focusing on the promotion and regulation of innovation.

I shall be exploring these issues in much greater detail in the next chapter. But they are the background to the increasing pressure to produce science which is 'exploitable', to break the barriers between 'basic' and 'applied' science, to target limited resources towards 'strategic' areas, and so on. We have explored four key aspects of contemporary science policy associated with this trend: the attention given to technology transfer, the use of expertise as a source of authority and reassurance over controversial science, the desire to measure the productivity of R&D through the development of quantitative measures of science output, and the need to review this output and endeavour to make it accountable to a wider public.

Drawing on a variety of sociological research, I have suggested how these features of science policy are much more problematic than at first appears. We saw that the transfer of technology is no straightforward matter since what 'the technology' actually is and hence its use in new contexts are matters for continued negotiation and debate between different interest groups involved. Here we can see the value of the social constructionist perspective within the sociology of science that was outlined in Chapter 2. Policies that ignore this negotiated and constructed sense of viable and successful technologies by treating technology as a black box will be less sensitive to complex social process through which technologies eventually emerge. As such they may be less conducive to technology transfer than their architects hope.

We have also seen how expertise is socially contigent, that it is informed by different interest-positions and that its role is often ritualistic in major controversies over science and technology. The interests approach discussed in Chapter 2 can then be brought to bear on this issue, as Nelkin (1974) and Wynne (1982) among many other sociologists have shown. Collingridge and Reeve's claim that expertise *should not* play a directive role in policymaking is arguable in both practice and principle, and conflicts with the critical position advocated by Chubin, Fuhrman and Shapin discussed at the end of the previous chapter. Clearly, if expertise is informed by different interests operating at differing levels and drawing on different sources of support as it moves into an increasingly wider and more public forum, scientists will endeavour to anticipate the claims and actions of other science groups (other 'experts') seeking to shape science policy debate. As Shapin has suggested, therefore, sociologists should apply a 'calculative model' to scientists as social actors. This would allow the possibility that, against the Collingridge and Reeve position, experts do not 'cancel each other out', but instead some calculate better than others and so *do* have a greater influence on policymaking than others. I so, it would be possible in principle for the sociologist to show that when 'science speaks to power' (as Collingridge and Reeve's text is called) some are listened to more favourably than others. Given this, it would be possible to offer not only a critique of the content and mechanisms for making science policy but also *how it might be otherwise*, the ethical role of sociological inquiry that Chubin suggests.

This point is particularly important for another of the aspects of science policy dealt with in this chapter, the regulation of innovation and existing technological systems (such as petrochemical plants). Irwin and Vergragt's work shows how the regulatory process needs to be regarded as a much more complex phenomenon than it often is. One can see how development in the sociology of science have informed their work, most notably, the stress on the *socially negotiated* character of regulation as one explores the perception of an response to regulatory constraints within and between different organisations. In showing the interactive and negotiated character of these processes, their aim is not to devalue regulation *per se*: on the contrary, they argue that their analysis encourages a more refined approach to the development of regulatory frameworks which can meet the needs of both those subject to regulation and those whom the regulation protects.

As a result of the sociological research reviewed here, Ronayne's picture of science policy as 'authoritative and informed decision-making' is perhaps too glib a portrait of what is a much messier affair. What we can see, however, is that the complexities of the picture have been revealed by sociological research that has developed on three fronts over the past two decades. First, the developments in the *sociology of science* charted in Chapters 1 and 2, then the emergence of a *sociology of science policy* sketched out through the research presented in this chapter, and finally, drawing on these two, the first stirring of a *sociology for science policy* which I have hinted at towards the close of this chapter.

We have to recognise, however, that the sociology of science and science policy is itself just one of the many 'calculative' voices heard in the decision-making arena. Hence, just as we noted earlier that scientific interests may be unevenly represented and heard in this arena, so too with sociology: we should not expect to have an increasing impact on policymaking simply because research and analysis grows increasingly sophisticated and synthetic. There will be a whole host of contingent factors that will determine whether the government is interested in what one has to say, including the perceived 'timeliness' of the work and the allies one can call on based in government agencies. Even if both of these obtain, it is not obvious that one's research will have an impact on government: as Gummett (1986) has said:

> The real test of policy relevance is not who paid for the work, because although having a governmental customer may help; it does not guarantee an impact on policy, and there may in addition be other reasons for government support than curiosity as to the results. The real test is rather, the question, was there an impact? Here, however, we enter somewhat difficult territory. We have to consider what it means to talk of a policy impact, and how we would detect one (p.61).

To do this, we need to understand the priorities and practices that underlie science policy, and how these vary from one state to another. This is the focus of Chapter 4.

4 Exploiting Science and Technology (I)

Introduction: political culture and science policy

The theme of the next two linked chapters is the exploitation of science and technology. In this chapter the primary focus will be on the ways in which different states have sought to exploit their national knowledge bases of behalf of the 'national interest'. We shall also give some consideration to the way in which this is increasingly associated with pressures to commercialise public sector R&D, whether in the laboratory of the government research establishment or the academic institution. As we shall see, this commercialisation has also been accompanied by a restructuring of public sector research. Although state-funded research conducted in the public sector is very important – especially in the military field – corporate R&D is much greater in terms of funding and personnel. This industrial exploitation of science and technology is discussed in detail in the next chapter.

The way in which governments exploit scientific research varies quite considerably, not only in terms of what is exploited but also how this is done. This variation reflects the unevenness of different countries' science bases as well as differences in their more general *political cultures*. Policies for science are, then, only properly understandable in terms of the political context in which they are fashioned. Economic factors play their part as well, of course – as we saw, for example, in the case of the 'failure' to privatise the British nuclear power industry. These factors will be explored more fully in the next chapter, however, for sake of clarity.

The phrase, 'political culture', refers to forms of political behaviour, ideology and institutional structure that characterise the way a state and its citizenry exercise power. Different political cultures have different political arenas, to which interest groups – feminists, scientists, unions, civil rights organisations, corporate capital, or whatever – have different and unequal access. Such arenas also differ in the way decisions are made and the degree to which they are accountable to a wider 'public'.

Science policymaking is only one of many activities conducted within the political arena, and must compete against other considerations of the state for resources and power. So, science policymaking agencies will operate in different political cultures and enjoy more or less power within the state. Before I look at the variety of science policy in industrialised states, I can illustrate the way political cultures and arenas vary by considering briefly the differences between the United States and Britain in this regard.

The United States is often described as a pluralistic democracy in the sense that competing interest groups can make their respective voices heard on the political stage via the representative institutions available. The style of political action can be said to be gladiatorial – an open, competitive struggle between different groups. This tends to encourage pressures for accountability to be guaranteed through formal constitutionally-based laws and statutes that enshrine civil rights as well as obligations.

While this image of American democracy is often overblown – especially in the media and the rhetoric of political leaders needing to justify yet another invasion in Central America – there is some institutional substance to it. Statutes requiring release of an enabling access to information do exist – such as the Freedom of Information Act – and have been used by the less-advantaged or pressure groups to challenge the power of the political and economic elites. Of course, those who enjoy considerable economic and political advantage may endeavour to develop countervailing statutes that limit the effect of any challenge from below.

In the British context the situation is somewhat different. British political culture is characterised by a discrete exercise of power through the network of the conservative British 'establishment' (Wiener, 1981; Scott, 1982), keen to sustain its economic and political privileges in 'this green and pleasant land'. The brash, hectoring politics of America has no genuine counterpart here, apart from the ritualistic exchanges on the floor of the House of Commons. Kinship links and social networks are the more traditional vehicles for a discrete bit of establishment politicking: as Scott notes:

> the establishment emerged as an important social and political force during the nineteenth century and has been seen as a central element in the 'antique' or 'patrician' character of the British state. . . . [I]t facilitates communication amongst those who are familiar with one another, share a common background and meet in numerous formal and informal contexts (pp.158–9).

This is not to say that the groups that make up the British elite are always in agreement: at times of economic or political crisis their unity is more likely than during less critical periods when, for example, changes in state policy may hurt some more than others.

These broad differences, only briefly and perhaps too glibly sketched out here, can be detected in the way in which science policy is made. Thus, Jasanoff (1989) contrasts the US and British science policy *cultures* in the following terms:

> In the US, major science policy decisions are virtually certain to undergo challenge in court, despite complaints by scientists that judges should not have the power to 'second guess' administrative agencies or to decide disputes over technical evidence. Further, . . . [there is] an adversarial flavour to relations among the major players in policy making, so that the US administrative process seems to advance more through a series of formal offensives and counter offensives than through negotiation and compromise. . . . [However, in the British and, more generally European context] political interests with a stake in science policy appear to be more accustomed to ironing out their differences without open conflict (p.368).

This distinction in transatlantic policymaking has been explored empirically in a number of comparative sociological studies (see e.g. Vogel, 1986; Brickman et al., 1985). One of these, by Gillespie *et al.* (1982) shows how the distinct political contexts of US an British science policy lead to different decisions being taken over public interest issues, in this case about the hazardous nature of two chemical pesticides, Aldrin and Dieldrin. Again, like Jasanoff, Gillespie draws our attention to the broad differences in political decision-making:

> In contrast to the conflicts among experts that characterise many American decisions in [the area] British decisions emerge from a closed decision-making process . . . Whereas US decision-making institutions depend upon, and to some extent, generate conflicts among experts, British universities tend to rely upon singular sources of expertise (p.329).

The adversarial character of the US science policy process led ultimately to a ban on the two chemicals because of their association with cancer, whereas the British authorities decided against a ban, even though the experts on both sides of the Atlantic 'reviewed the same experimental evidence of [the chemicals'] possible carcinogenicity' (p.306). Gillespie recognises that, as we have suggested at various places elsewhere, the interpretation of the evidence could have been different, and acknowledges that there were uncertainties among both US and British experts. However, while these uncertainties were important, their effect depended on the *political rather than the technical context* in which they were expressed. It was this that determined how the evidence and its ambiguities would be dealt with.

Studies such as Gillespie's indicate that very different decisions may be taken about science policy issues as political contexts and cultures change, despite broadly similar attitudes towards the evidence. The US and British political contexts can be contrasted in terms of their relative degrees of openness to a wider public influence. This is one useful way of characterising science policy-making. Another is in terms of the degree to which decisions are part of a wider national framework, a national policy towards science planning by the government and its state agencies.

Some countries may have a loose, pluralistic structure through which the planning process for science is developed. Others may adopt a much more centralised, directive style, from what might be called a 'corporatist' approach where state, industry and academia, along with different public representative groups work together to fashion a programme for science, to the strong state directive programmes, such as the USSR's State Committee for Science and

Technology, until recently a top-heavy, over-bureaucratic institution, now reformed through Gorbachev's strategy of *perestroika*. Gorbachev has also sought to increase the linkage between researchers and industry through the establishment of 'interbranch scientific and technical complexes'.

Taken together, these two key dimensions of science policy produce a convenient, if not highly sophisticated, framework for describing many countries' approaches towards science policy according to the political culture they have. A simple matrix diagram can be produced (see Table 4.1) within which specific countries can be located. A number have been selected for purposes of illustration. By no means can this simple fourfold matrix model capture the complexity of science policymaking within and between countries, but it does offer a useful first approximation to the broad dimensions by which they can be distinguished. One has to explore countries more carefully to see what exactly, for example, 'state-planned' actually means. National planning for science and technology can take various forms and be managed by more or less centralised, more or less accountable bureaucracies. Different states may also change the style of policymaking: for example, in 1989 the US Congress debated the establishment of a coherent national policy for science to coordinate the exploitation of new technologies in face of increasing competition from Japan (Swinbanks, 1989). In Britain too, there are increasing calls for the adoption of national programmes in certain fields, such as a National Biotechnology Programme to orchestrate work that cuts across different research centres. Typically, the more centralised the policy the more this is likely to mean that government accepts responsibility for *both* 'basic' and 'applied' R&D, that is, develops policies that promote new technologies as well as encouraging an

Table 4.1 *Political culture and the characterisation of science policy*

	Pluralistic	*State-planned*
Open/competitive	USA	Netherlands
Closed/limited public scrutiny	UK	Japan

industrial infrastructure which will exploit them, in short, policies that bridge both technology and industry.

In the UK, the two principal government departments responsible for science and technology, the Department of Trade and Industry (DTI) and the Department of Education and Science (DES) have been prepared to initiate policies for technology but not industry: during the Thatcher administrations, an increasingly arm's-length approach towards industry was adopted on the assumption that the private sector would be the best judge of new innovation and its market potential. This approach can, however, lead to difficulties if industrial capital is slow to restructure and take advantage of new technologies. For example, this has happened in the case of the British information technology industry. The government's support for new technology, especially via the Alvey programme (see Chapter 3), has not been accompanied, quite deliberately, by any matching industrial policy for the IT industry. Guy (1987, p.101) has commented on this lack of policy downstream for incorporating the new technologies in a major way: 'spin-offs' have occurred, but these have been insufficient to stimulate renewed, extensive growth in the sector. The government's policy has been to encourage the development of so-called 'pre-competitive' research – that is, research which is focused on the basic/strategic rather than developmental end of innovation – through its LINK programmes in areas such as biotechnology and advanced composite materials science. These bring together academic, industrial and governmental groups, with matching funding from industry and the state.

The British government's policy towards the development of science and technology can be best described as a 'supply side' approach in the sense that it helps sustain innovation upstream while placing the responsibility on the demand side, broadly speaking the industrial sector, to pick up and exploit the new ideas that appear in what might be called the 'R&D marketplace'. Though developed most fully during the Thatcher era, this strategy can be traced back to 1971 with the appearance of the Rothschild Report (1971): this was the first significant step in creating what Clark (1985) has called the 'social facsimile of a market situation – when consumers [e.g. industry] 'purchase' the research they need from contractors [e.g. government research labs or universities]

Table 4.2 *An example of what an extensive and comprehensive national material conservation policy would look like*

Type of measure	Target area	
	Ambience	*Industry*
Financial	Cheap long-term loans, expensive short-term money	Support for materials saving design; support for recycling; research grants for materials conservation
Taxation	Lowering of relative cost of labour vs capital	High tax on new materials, lower rates on recycled materials; similarly, high tax on new machinery, low on reconditioned
Legal and regulatory		Control of waste materials; controls on writing-off of machinery
Educational	'Old is beautiful' education	Materials engineering courses
Procurement		Minimum arms purchases; specifications written to save materials
Information	'Waste not, want not' campaign	Recycling and materials-saving information centres
Public enterprise		Research on recycling, on materials-substitution and on materials-saving design
Political	Peace in our time – for war is the greatest waste of materials	
Scientific and technical		Research institutes for materials conservation; support services for materials conservation and substitution

Source: E. Braun, *Wayward Technology* (1984)

Commerce	Foreign trade	Consumers
Incentives for collection of used materials	Tariffs on selected materials	Incentives for recycling; support for repairs to make goods last longer; long-term loans for consumer durables
Tax incentives on recycling operations and material collecting services		Decreasing tax rates for older durables, including cars
	Import restrictions on selected materials	
		Propaganda on value of materials; 'old and working well is as good as new'
Public trading companies for materials		Nationwide repair services; material collection services
	Re-negotiation of trade agreements	

who are supposed to provide it' (p.227). However, as I shall show later in this chapter, the 'contractors' had a rather lean time of it during the 1980s as the British government sought to restructure and cut back on research activity in the public sector. This brings us to the general question of how much do governments spend on their science and technology base, and where does the money go?

State funding for science and technology

Most countries within the Organisation for Economic Corporation and Development (OECD) have steadily increased their support for science, apart from a prolonged period during the 1970s when both economically and politically industrial capitalism experienced something of a structural crisis. In Britain, the problems have lasted much longer, and the government responded during the 1980s with a squeeze on public expenditure, which among other things, hit education and science. While other OECD states began to experience a genuine growth in their science budgets during the 1980s, the same cannot be said of Britain: there will continue to be a real cut in expenditure on R&D by the state over the coming years.

As with any statistical series, international comparisons of spending on science need to be treated with caution as they may not compare like with like: the way statistics are collected varies from country to country, the definitions of R&D may not be constant, and the relative buying-power of state support may mean that a dollar, pound or yen will buy more in one country than another. With these caveats in mind, Figure 4.1 summarises the latest information on state support for science in OECD countries for different areas for 1980 and 1987.

Generally speaking, most OECD states spend between 2 per cent–3 per cent of their Gross Domestic Product on support for R&D. Given the cutbacks in UK funding, however, British science and technology spending has fallen in both relative and absolute terms compared with its major competitors. While the latest government forecasts for expenditure on science (see Figure 4.2) involve a gradual increase in cash terms, in real terms (given inflation) the budget will decline in value by just over 10 per cent between 1987 and 1992, the cuts falling heaviest on the Universities Funding Council (UFC) and government's departmental research effort ('Civil departments').

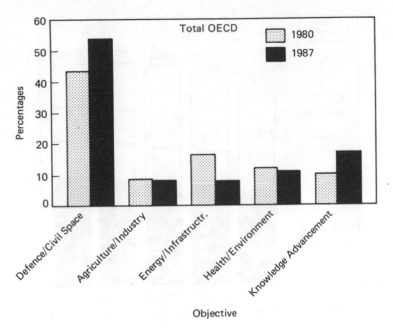

Fig. 4.1 *Government R&D budgets (by socio-economic objective: percent of total)*
Source: OECD, *Industrial Policy in OECD Countries*, Annual Review, 1989.

A notable feature of Figure 4.2 is the proportion of expenditure going to military R&D: this amounts to about 50 per cent (in 1990 the figure was £2.54 billion) of total R&D funding, a figure only marginally higher in the United States but much in excess of other OECD countries, especially Germany and Japan (with figures of 15 per cent and 13 per cent respectively). Very little (less than 20 per cent) of the research conducted within the Ministry of Defence will produce technology spinoff with any civilian value (ACOST, 1989) despite the MoD's commercialisation agency, Defence Technology Enterprises, established in 1985. Much of the money goes to fund contracts placed in the private sector to pay for the development or enhancement of weapons systems, such as the joint British/West German European Fighter Aircraft Programme costing an estimated £20 billion. Many British scientists have criticised the scale of support given to MoD research while more basic research has been cut back elsewhere. In fact, there are changes taking place in

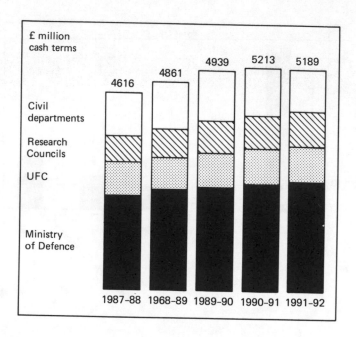

Fig. 4.2 *UK government-funded R&D (£m) 1987–92*
Source: Annual Review of Government Funded Research and
Development, 1989.

the way Britain and other NATO countries organise their defence
R&D, most notably in terms of a growth in *trans*national systems
of weapons procurement within which defence companies (such as
Ferranti or British Aerospace) are being asked to play rather
different roles compared with the past: as Gummett and Walker
(1989) have noted, today 'the responsibilities vested in firms for
defining technical responses to operational requirements are
greater; and so is the onus upon industry to fund more of the
underpinning R&D' (p.198). This notable concentration of funding
in one area is not, however, exclusive to Britain: while Japan has
very little military R&D, almost two-thirds of the budget of its
Science and Technology Agency goes towards R&D in nuclear
energy.

The general rise in expenditure on science and technology within
(most) of the OECD countries over the past decade has been

accompanied by a number of changes in the way this money has been used. These changes include the following:

(i) an increasing *selectivity* in the choice and direction of R&D in response to both perceived needs and opportunities for techno-logical innovation: this has meant a growth in 'strategic' research and an expansion of mechanisms whereby the technologies it generates can be transferred to industry;

(ii) a *concentration* of funding to develop or sustain a critical mass of scientists, often combining interdisciplinary skills;

(iii) an increasing support given to *programme* rather than indi-vidual project research activities (in the UK, for example, this now approaches 25 per cent of all Research Council funding): this accentuates the trend towards large research teams net-working across a number of laboratories;

(iv) a *tying of public sector funds to matched industrial expenditure* on R&D, with the result that commercial sponsorship of acade-mic R&D is growing steadily: it has also grown *within* industrial sectors, too, such that for a country as a whole industrial R&D funding may be greater than that of the state. For example, in West Germany, of total R&D, 39 per cent is funded by government, 61 per cent by industry; in Japan, the gap is even greater at 20 per cent and 80 per cent respectively;

(v) Finally, a growing *transnational research effort* with funding from different countries pooled in support of continental or global research programmes managed by major science policy offices, such as the European Commission's BRIDGE, SAST, and MAST initiatives.

These changes have been controversial. Many, rehearsing the professional rhetoric of their occupation, see them as challenging the 'autonomy' of science, while others regard the growing com-mercialisation of public sector science as a threat to curiosity-driven research. The advent of selectivity, concentration and managed programme research has also been accompanied by an increasing demand from sponsoring agencies that scientists and technologists give 'value for money'. We noted in the previous chapter how the development of quantitative and qualitative indicators to measure research output is likely to make an increasing contribution to policymakers' evaluation of scientific productivity.

At the same time, the data – such as the science citation index – that have been used to measure productivity and quality of research, can also be used to provide policymakers with an indication of the emergence of new areas of inquiry developing within scientific fields. Sociologists have been developing various techniques for monitoring the growth of new specialties, especially mapping out co-citation networks among scientists. Rothman (1984), for example, has suggested ways in which the mapping of scientific activity within particular fields can identify 'a change in intellectual focus' and so chart the changing pattern and progress of scientific specialties (see the boxed text). Callon (1979) advocates a similar programme of strategic mapping in order to understand the advent of new 'research fronts, their evolutions and interactions' (pp. 1–2).

Rothman has shown how one can create a model of an existing or emerging science speciality through examining the way scientists cite each other's work (co-citation) to suggest patterns of common intellectual interest. His initial method for doing this is reproduced here.

Science mapping for strategic planning:

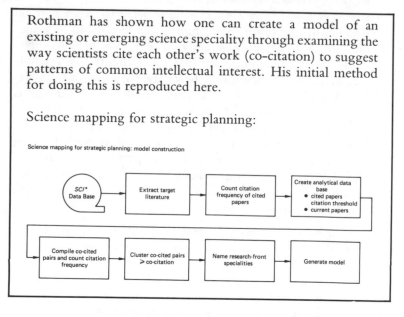

Science mapping for strategic planning: model construction

It is quite likely that both these evaluative and mapping techniques will become increasingly refined. Not only do governments press for this information in order to plan strategically for science (see, for example, Smith *et al.*, 1986), but corporate capital too seeks information of this sort in order to discriminate between research which has potential commercial value in the technological arena within which they operate. Knowledge and information *per se* are certainly vital factors of production for contemporary capital but *structured* information has more added value and reduces the search through the mountain of scientific literature that grows at an annual rate of over a million publications each year.

The trends in science policy identified above have meant that the science base is being managed in such a way as to promote, on the one hand, a rapid specialisation of knowledge tied to certain institutional centres of 'expertise', while on the other, cross-institutional links that blur cognitive boundaries between disciplines. In this context, the traditional image of science as individual, autonomous research – always something of a myth as Latour (1985) and others have shown (see Chapter 2) – has become even more difficult to sustain. The commercialisation, restructuring and managed character of contemporary science has meant that scientists, as 'calculative actors', have to be predisposed to enter the policymaking and economic arenas outside of their immediate research network much more routinely than in the past.

Since these trends in policy can be found in both decentralised, pluralistic as well as more centrally-planned policy contexts, scientists in both will have to be as 'calculative' as each other, though no doubt using rather different means to sustain their authority and their professional practice. Scientists developing new areas of research – such as genetic engineering or superconductivity – will have to invest their scientific capital according to the sort of political culture within which they find themselves and the sort of economic terrain on which they are seeking to implant new ideas. As we have seen in Chapter 3, the translation of a set of ideas or techniques into something that becomes an established part of what we might call the technological infrastructure of society such that it is embodied in products or production methods is not straightforward. Economically, much depends on the way the new technology is perceived to impinge upon existing technological systems: public or private sector industries may accommodate it without major changes being necessary, may adapt to it at the margins, or may adopt it as a 'revolutionary' technology that transforms the productive process as a whole. As Pavitt (1983) has shown, there is no simple or single pattern to the process of innovation.

Not only is this true in broad economic terms, it is also true spatially. The diffusion of technological innovation is an uneven process reflecting political and economic inequalities and opportunities within and between different countries. Advanced industrial capitalist states have a more broad-based and developed technological infrastructure than say Second or Third World states. Yet even here, spatially unevenness in this infrastructure is apparent. This in part explains the growth of *regional* science and technology policies within OECD countries. I shall now consider some of these policies, including the most fashionable but probably most 'hyped' of all, the development of 'science parks'.

Regional policies for science and technology

From the late 1970s onwards, large multinational corporations undertook an extensive programme of restructuring in response to a chronic downturn in profitability. Among many other changes, this involved a growth in un- and under-employment in mature capitalist states as production moved to areas (especially in the global 'periphery' – the so-called Third World) where labour was relatively cheap. Production was no longer to be geared to large-scale manufacturing plants located in the industrial heartlands of the capitalist centre, but to be spread, made more flexible, and more technologically sophisticated, 'smart', or 'hi-tech'.

What were once centres of manufacturing became centres of depression. By the end of the 1980s, the old industrial centres had responded in a number of ways. Often backed by central government, regional authorities, in many OECD states, have tried to develop new employment opportunities through local development schemes based on new technologies, sometimes organised on a regional basis in technology networks. These have been notoriously difficult to finance and sustain, and have in fact generated relatively few new long-term jobs. The new growth areas – the Silicon Valleys or Glens – of this world grew by a strong combination of a regionally rich concentration of innovative technological expertise and significant defence-related R&D for the state. Indeed, it appears that many of the growth areas in the UK located in the South and South East have been heavily dependent on defence contracting to sustain the levels of investment they have enjoyed. As Simmie and James (1986) noted, 'In fact, many parts of the south east would be joining the north as depressed areas if they had to rely on the commercial, demand-pulled civilian electronics industry as their main source of industrial growth. It is defence expenditure that makes the difference' (*New Society*, 31 January). Indeed the contribution of defence R&D to UK corporate success should not come as a surprise given, as we saw earlier the level of support it receives from the government. The boxed extract from the *Guardian* gives an indication of how important defence contracting is for the major UK companies.

Market's nuclear activity fallout

Almost 50 quoted British companies are involved in major nuclear weapons activity, according to research by the ethnical investment service Eiris. Excluding these companies from an investment portfolio would remove 18% of the stock market, measured by market capitalisation.

Eiris investigated companies involved in the construction of nuclear bases, those working on Star Wars research as well as defence contractors for nuclear weapons systems . . . Twenty six companies are working on nuclear weapons systems or components, the ten most important receiving more than £100 million from the Ministry of Defence in 1985/6 . . . Excluding all these shares would still leave 82.5 per cent of the stock market. Cutting out conventional as well as nuclear armaments suppliers would only leave 43 per cent of the stock market for investment . . .

Source: *Guardian*, 12 April 1988

There are, then, major differences between regions in terms of the levels of both public and private investment they enjoy. In less prosperous areas, the state may take a more or less direct role in redeveloping the local economy. It is much more likely to establish formal agencies of state to do this where national programmes for both technology and industrial policies exist. Elsewhere, the central state may rely on, and perhaps indirectly support and work through, local bodies such as regional or metropolitan councils or the individual federal states in the US or West Germany (the eleven *Lander*). The latter have been particularly important in encouraging industrial capitalist growth precisely because of their formal legal and executive powers to finance and steer local initiatives independently of central government. To the extent that such middle-range institutional authorities were emasculated or even entirely removed in Britain during the early Thatcher administrations, the possibilities for effective innovation policies in the older urban areas outside of the South East became that much more doubtful.

In the US, however, local state government was encouraged during the same period when Reagan was in power, something of an irony, given the supposed Reagan–Thatcher ideological rapport. Cowan and Buttel (1987) suggest that the Reagan period saw the emergence of 'regional-sectoral corporatism'. This 'is characterised by decentralised cooperative arrangements among state governments (or regional governmental agencies), universities, and small-to medium-sized, regionally-based "high-technology" industries which may or may not be tied directly to large corporations' (p.4). They argue that the Federal state uses these more localised technology innovation networks in order 'to produce and diffuse new science-based technologies capable of restoring US economic competitiveness' (p.8).

Powers *et al.* (1988) have provided a detailed commentary on many of these state-sponsored initiatives in high technology and show the importance of the local fiscal policies – such as 'tax holidays' on R&D investment – in attracting new business into their regions. Similar regional support programmes can be found across Europe: in the Netherlands, the government has recently established 18 regional Innovation Centres through local authority offices; in West Germany regional states have established since 1985 a network of 12 technology centres to try to respond to the decline of traditional heavy industries; in France, the regional authorities have established 'technopoles', local high-tech industrial estates. Clearly, all such policies are new forms of state support for industrial capital at a time, during the 1980s, when capitalist interests were restructuring the manufacturing base in order to participate in the new technologies on which their competitive future depends. The long-term success of multinational corporations will depend as much on their technological as it will on their trade advantages over other multinationals.

However, these state policies have often had fairly limited effect in bringing about the rebirth of declining regions and promoting the leaner, high-tech capitalism of the twenty-first century. As Cowan and Buttel have shown for the US and Marvin (1987) for the UK, typically regional (and thereby, by implication, national) policies for industrial regeneration have lacked sufficient power to restructure local and national industries, especially in the declining, or 'sunset' industrial regions. They contrast this situation with Japan where, under the powerful direction of the Ministry of International Trade and Industry (MITI) new nationally-orchestrated science policy programmes in IT, biotechnology and other areas have been established in cooperation with a receptive industrial base. The success of the Japanese lies, therefore, in a strong state acting in a corporatist way with industry. This situation has not arisen overnight but can be traced back to the latter part of the nineteenth century when a strong, reformist and modernising imperial government set about the radical modernisation of the country's productive base. This culture of strong 'statism' has been seen since 1945 in a whole range of *industrial and not merely technology* policies for Japan. In much (but not all) of Western Europe and the United States, the general strategy has been to develop regional and national support for new technology without the accompanying industrial programmes to translate invention into innovation. As Cowan and Buttel comment, 'Innovation policy is a hybrid of wedding science and technology policies with industrial policy instruments' (p.19). Industrial policy is needed to push an often reluctant private sector to move into new technological territory, as Rothwell and Dodgson (1989) have shown.

While regional technology policies have many common features given their broad aim to promote local development, they can, however, vary considerably in the way they (falteringly) try to achieve this. The sort of policies chosen reflect the sort of assumptions made about the best strategy for economic growth and the kind of mechanisms available that can be used to implement new programmes. Marvin (1987) has shown, for example, that in the UK two very distinct types of regional policies have appeared which reflect different political and economic assumptions about the best way to achieve growth based on new technologies. Most have adopted a 'market-led' approach to local technological development. While this has helped foster some growth, especially in the form of 'science' or 'technology parks', it tends to be very limited in both its geographical spread and its creation of new jobs. A few, such as Sheffield and the former Greater London Council, have started from a different position, trying to stimulate local technological innovation and growth through policies based on meeting various social needs' most importantly, employment. 'Need-led innovation will usually create jobs as the aim is to create new products and provide extra services rather than increase productivity' (p.109).

Despite the fact that market-led development has generated little in the way of new technologically advanced infrastructure for British industry as it approaches the year 2000, it has been in tune with the political ideologies of the 1980s and so received much support. Perhaps the glossy image of the 'science parks', more than anything else, epitomises this faith in smart, entrepreneurial dynamism.

UK science parks

The science park image is a collage of designer science and GTI-technology fashioned amid PC terminals, polished aluminium and glossy potted plants, a high-pressured environment and no place for the faint-hearted. Located on the outskirts of a town or in a rehabilitated old industrial area, the green, clean feel of the park is far removed from the smokestack images of older industrial production. A science park address adds kudos to the starter company looking for new customers. At the beginning of 1990 there were 38 operating parks with over 800 tenant companies. There are, however, a few parks which are significantly bigger than most of the rest. The largest are the first two that were established, Heriot-Watt and Cambridge in 1972 and 1973 respectively. The parks have often been said to be a response to the Labour government's 1964 call for universities to help create companies on the frontiers of knowledge (Wright, 1985). They are seen as the UK version of the 'Silicon Valley' phenomenon which provided a model for the development of European and American multinational companies (*The Economist*, 1985). According to the United Kingdom Science Park Association, science parks must:

(i)   develop formal links with a university or other public sector research centres;
(ii) promote the growth of research-intensive companies;
(iii) endeavour to transfer new technology to the commercial marketplace.

There have been a number of detailed studies of UK science parks. Lowe (1985) and Currie (1987) provided very similar assessments of them, concluding that, apart from one or two exceptions, they have had a fairly limited effect on innovation rates and the promotion of academic/industry links despite their image to the contrary. Lowe comments that while most of the focus of the parks has been on high-tech businesses, this may not be 'the major innovation problem faced by industry: better design, quality control and the adoption of already known and tried technologies, may be pertinent to the needs of British firms' (p.32). And even in the high-tech context, location on a science park 'may be neither a sufficient nor a necessary factor for high-tech company growth' (p.37).

The parks have appeared in two stages, reflecting the different circumstances in which they were formed. The first few in the early 1970s mimicked the development of a group of innovating firms around key academic centres, such as the Massachusetts Institute of Technology. Often there were strong links into defence technology which helped provide a secure financial base for the new high-tech firms. The now classic study of the science park, Segal Quince's *Cambridge Phenomenon* (1985), points to the importance of the general growth in and around Cambridge of advanced electronics firms by local and multinational industry and Cambridge academics among whom strong but informal links have developed. The park itself, situated just to the north of the city, has (in 1990) 81 tenants and average annual rents of £15 per square foot which can often deter the smaller start-up company from locating there. The site for the park is owned by Trinity College, which initiated the park with property development considerations uppermost in mind, currently renting out almost 600 000 square feet, making it the largest UK science park, twice the size of the next biggest, Heriot-Watt.

The second group of parks appeared during the 1980s and were much more to do with universities' response to government expenditure cuts in higher education. As Wield (1986) notes, these cuts (in some institutions over 40 percent [e.g. Salford]), forced universities to find other ways of raising income, including:

attempts to use surplus land to attract high technology companies; attempts to find industrial, bank, and other local government sponsorship for their applied activities including science parks; attempts to improve their image vis-a-vis a 'privatising' government by being seen to attract private capital to commercialise their R&D (Wield, 1989, p.67).

At the same time, local authority support for science parks added to the momentum during the 1980s. Much of this derived from the belief that the parks would help overcome the rapid rise in unemployment seen during the decade. Despite these hopes, the parks have had very limited impact on areas of high unemployment, and where they have been successful in generating work both on and off the park (as in Cambridge) this has been because of a more favourable pattern of industrial growth in the wider regional context. Science parks are clearly not, therefore, the solution to long-term structural depression and unemployment.

It is also important to recognise the *sort* of work and work experience that the parks have generated. The parks are primarily R&D, consultancy centres with limited production installations on site. Employees are typically graduates or postgraduates working in a flexible managerial context that gives considerable autonomy and usually high pay, but in a high-pressured, uncertain environment. The occupational ideology of the parks dismisses work collectivism as a thing of the past, irrelevant to the progressive, individualism of the modern era, the era of the 'new silicon idealism', as Robbins and Webster (1988) put it. Massey (1985) has provided a rich picture of science park culture:

> Within this idyllic setting the organisation of work could not be more different from, and is explicitly counterposed to, the conditions that most of the population daily faces in office, in mine, or on the shop floor. In this classic, high status science park, the social relations of work were characterised by individual flexibility between jobs; individual commitment to the company; frequently decentralised management structures; clean bright carpeted workplaces; individual interest in the job; voluntary flexibility in hours of work (image: staying on into the night, struggling over the knotty problem on the frontiers of science; and a total absence of trade unionism. This last is no accident. The whole construction is one that emphasises individuality . . . as opposed to collectivity and solidarity (p.310).

Indeed, in order to ensure that this individualistic ideology prevails, one finds management consultants advocating quite explicitly that science parks should be established away from areas that have had a history of industrial collectivism, as the following US commentary shows:

> A close look at Silicon Valley or Cambridge Science Park reveals that an area with a tradition emphasising individualism will likely prove more successful for a science park development than will an area with a long labour union history, with its emphasis on majority opinion and maximum group participation . . . Areas targeted for economic development initiatives should be selected to include this cultural characteristic (Russell and Moss, 1989, pp.60–1).

So much for the value of 'majority opinion' and 'group participation'! In fact, as Wield (1986) has argued, the evidence that is available derived from both science parks and elsewhere, indicates that the high-pressured, uncertain work environment of the science park employee is unlikely to promote higher productivity or efficiency, and may reduce the level of genuinely innovative activity.

Despite their image, therefore, the UK science parks have had only limited impact on employment patterns and the country's broader technology base. They are unlikely to be the catalyst through which the British workforce will become more attuned to the new technologies being developed. They will probably be principally of use to the large multinational companies that will keep a 'window' on the park, by owning a subsidiary there or developing licensing agreements with science-park companies to develop and commercialise high-tech products and processes. To the extent that they do this, local governments that support parks may find they indirectly damage more traditional industries in the area, and so threaten a renewed growth in unemployment. Clearly, though, whatever their local effect, science parks are related to some of the trends in the institutional character of science and technology outlined in previous chapters. They reflect both increasing state intervention at the local level in the mnagement of science as well as the increasing shift towards the commercialisation of knowledge bases within academia. It is to this and the wide-spread restructuring of public sector science that I shall now turn to complete the first part of our examination of the exploitation of science.

Restructuring and the commercialisation of public sector research

If one could select two terms that might best characterise changes in the organisational or institutional basis of public sector scientific R&D over the past decade, those of 'restructuring' and 'commercialisation' would probably be most appropriate. Within the science–policy context they are directly tied into the four processes marking out new institutional directions for science and technology discussed in Chapter 1. They are also linked, inasmuch as developments in restructuring the science base feed back on processes associated with its commercialisation, and vice versa.

The *restructuring* of public sector science within academic and government research establishments has been part of a wider set of changes, in the planning, funding and management of educational and research activities throughout the OECD countries. Very little sociological analysis has been made of these changes, though there has been considerable, often heated, debate about them in and outside academic circles.

In the broader context, restructuring has been manifest in a number of ways:

(i) the cuts in state support during the early 1980s forced large-scale 'rationalisation' of academic departments, with redundancies, relocation of teaching and research staff, and demands for increasing 'productivity' and 'selectivity' in education;

(ii) the pressure to break down traditional disciplinary boundaries, again in terms of both teaching and research, and to develop new institutional frameworks for greater interdisciplinary work: in the research context within the UK, this has led to the establishment of new 'Interdisciplinary Research Centres' (IRCs); linked to this, has been the development of new types of strategic research programmes with industry;

(iii) the broad changes in the contractual relationships between public-sector academics and their employers, with an increasing emphasis on a more 'industrial', 'entrepreneurial' and 'efficient' pattern of work relations overseen by new forms of management – much more directive, much less 'hands-off': in the UK, associated changes here include the removal of tenure for university staff, the growth in short-term employment contracts for new and promoted staff and the incorporation of polytechnics and some larger colleges as independently-managed establishments free from local authority control.

There is a variety of interpretations available to explain the onset of restructuring. Clarke *et al.* (1984) suggest that it reflects the state's growing desire to control and direct educational research and training through new forms of corporate management designed to extract greater productivity from those working in the sector. Others, (e.g. Scott, 1984), highlight the sense in which these changes have had a particular resonance in Thatcherite Britain, challenging the cultural and occupational power and autonomy of the educational profession, thereby taming what has often been a source of independent critique of government policies. Both of these views have some merit and focus on the *political* and ideological dimension of restructuring.

At a more institutional level, and with specific reference to science itself, Ziman (1988) has seen restructuring as the inevitable result of new circumstances which characterise what he calls 'science in a steady state'. Arguing that these circumstances prevail internationally and are not simply associated with the parochial context of Thatcherite politics, Ziman points to the way governments in all countries are having to balance the demands for an enlargement of a more complex and expensive science base with the pressure for research to be more productive as well as accountable to a wider public constituency than it has been in the past:

> Science is thus moving into a dynamic 'steady state', in the sense that adjustments to change have to take place within a roughly constant envelope of resources, even though it is expected to serve the nation more efficiently and to account more directly for its costs (p.3).

Research becomes not only more collaborative but also more competitive, more sophisticated but also more expensive: all this has meant that resource allocation to science is managed in a more directive manner, both nationally and internationally. As a result, 'researchers have to be responsive to research priorities originating elsewhere in the political or economic environment' (Ziman, 1989, p.2). The notion of a 'steady state' might be criticised for offering too positive a gloss on various changes that have clearly had a very *un*steadying effect on those caught up in them: changing conditions of work, new forms of accountability, an increasingly uncertain and competitive resourcing environment, are all likely to have created difficulties for scientists. Nevertheless, to the extent that it points to significant institutional changes at the economic and normative levels, Ziman's model draws our attention to the sort of social demands and constraints scientists are having to face today: they must adjust their behaviour as *calculative actors* in response to these new circumstances. There has, however, been little or no detailed sociological exploration of how this is affecting the research culture of science within the public sector (but see Webster, 1990).

The related though analytically discrete issue of the *commercialisation* of public sector science has had more attention within the sociology of science and science policy fields. Commercialisation refers to a number of processes that encourages the introduction of market relations into the public sector institutions. Apart from encouraging a more 'entrepreneurial' enterprise orientation among academics, this has been done through establishing a range of mechanisms for linking academic individuals or research groups with R&D laboratories in industry. Accompanying the new enterprise culture is a whole host of organisational changes that mimic practice in the commercial context, including the appearance of corporate plans, full economic cost accounting, commercial law offices to handle intellectual property rights – especially patents – and the growth of small spinoff companies as well as large university companies geared to the commercial exploitation of their staff's expertise.

In the UK, this process was accelerated in part as a necessary response to the expenditure cuts of the 1980s; it was also given further momentum by the government's policy of pushing for greater industrial funding of strategic and applied research. Many of the government's own research institutes, especially those engaged in agricultural research, had to identify those R&D programmes that were 'near-market', and which should, therefore, be the responsibility of industry and not the Treasury. Though this device over 24 per cent of the budget of the Agricultural and Food Research Council was cut over the 1988–91 period, though industry did not, however, make good this reduction (see Read, 1989).

Government policy was enshrined in a number of Reports published during the 1980s (see Merrison, 1982; Muir Wood, 1983; Reece, 1986). The most detailed and analytical was the Mathias Report (1986) which, while accepting that more could be done to attract private investment into academia, noted that the declining profitability of UK industry in the mid-1980s would make this difficult to achieve. Rather than attempting to *replace* public funding for strategic science by income from other sources, Mathias urged that private and public funding must be 'complementary' and, crucially, that any revenue generated by commercial ventures within academia should be retained and not offset by an equivalent reduction in state grant. Despite such pleas, as we saw earlier in this chapter, the state has reduced in real terms its funding of research across the board – not merely in the applied arena. Industry has indeed increased its support for research within the public sector, though in broad terms this still amounts to a relatively small proportion of the total income the sector receives from all sources for its R&D activities: in 1987–88, almost 15 per cent of UK universities' total research grant and contract income came from industry (see Table 4.3).

Table 4.3 *Research income of universities 1983–88 (£m)*

	1983–4	1984–5	1985–6	1986–7	1987–8
Research councils	134	145	160	182	186
UK government	57	63	74	84	91
UK industry	32	47	59	68	78
UK charities/other	78	83	117	147	173

Throughout OECD countries, government policy over the past decade has been to 'lower the threshold to commercialisation' (Bullock, 1983) within public sector research establishments. This raises the broad question of *how far* can this sector develop its commercial links and activities without losing its essential character as *public* research, ostensibly open and freely available with its research priorities determined by intellectual rather than commercial interests? Simply because of the growth in industrial sponsorship for all forms of research (i.e. not just applied), should we presume that such sponsorship changes the range and type of scientists, questions, experiments, and conclusions? Clearly, one of the most important issues here is whether there are competing or different criteria used in the academic and industrial contexts, and whether the appearance of new industry/academia collaborations will lead to the development of new criteria, explicit and implicit, to assess scientific 'worth' and 'research productivity'. In such collaborations, the social shaping and construction of science is operating in a wider R&D marketplace compared with that discussed by Latour and other sociologists of science discussed in Chapter 2, with their focus primarily on scientists within academia. In fact, the sociological analysis of the commercialisation of scientific knowledge has its roots in a rather different tradition from that which Latour and other constructivists represent, one that can be traced back to the Marxist notion of capitalist industry exploiting scientific knowledge as a 'commodity' bought and sold as part of the 'valorisation' process of capital accumulation. We saw something of this debate in Chapter 1.

While the concept of science as a commodity acting as the principal value-added feature of today's productive process may be useful in formal terms, it is exceptionally difficult to operationalise the concept empirically. As Clark (1985) has argued, the value-added contribution of knowledge is only realised downstream in the productive process. Hence, it becomes 'impossible to place an unambiguous "value" on the scientific output itself, whatever its form, since the "value" of the final (economic) product is the result of a combination of many inputs of which scientific research is only one' (p.68).

Outside the specifically Marxist approach, there has been an attempt to refine our understanding of the *degree to which* knowledge plays an important part in the productive process of different types of manufacture. In other words, while all industries depend on 'knowledge', the degree to which their profitability relies upon it can vary significantly. Hence, a report by the OECD (1984) noted that there are 'many large firms active in fields in which technological change is less rapid, and in which market position depends upon factors other than level of technological sophistication,' citing shipbuilding and mining as two examples. Companies in these sectors are much less likely to have strong links with public sector research. However, such links will play an important role for high-tech knowledge-based firms in the new technology areas, such as biotechnology. Not only are these new, they are seen as *key* technologies for the regeneration of industrial capitalism as it begins a new 'long wave' of growth (Freeman, 1987), or as Marxian analysis would say, 'a new phase of capital accumulation'.

The high risks and development costs associated with innovation mean that large companies predominate and act as important 'knowledge bases' in the science system. Given high development costs – say for a new drug (currently quoted at about $125 million) – these firms will ensure that they enjoy as extensive a monopoly as is possible over their invention and innovation. Moreover, given the importance of technological advance to their wellbeing, the larger companies have to have access to the frontiers of knowledge and cover as much of the technological waterfront as is possible (Arnold and Guy, 1986). As Faulkner (1986) said in her study of the biotechnology waterfront, 'the requirement for strong linkage with academic research is likely to be an increasingly crucial element in the commercialisation of new science-based technologies (p.369). And as we saw, this pattern will become more common as the distinction between basic and applied research breaks down, as the experimental questions asked of data can pull in both basic and applied directions *at the same time* (see Wright, 1986; Webster, 1990).

Implications of commercialisation

The commercialisation of public sector science could be said to have three important effects:

(i) science might be said to be shaped or manipulated by commercial interests inasmuch as scientists find their research programmes being set by corporate sponsors;

(ii) the conditions of work, and the relationships between research teams may change;

(iii) free access to and the exchange of scientific information, data, materials and findings may be seriously compromised as the commercial secrecy required by companies limits the dissemination of ideas.

To what extent have these actually occurred? I will look at each in turn.

'As a commodity, science will be manipulated for their own ends by those who allocate resources for research' (Gibbons *et al.*, 1985). This statement summarises one of the basic theses of those who have explored the impact of commercialisation on public sector science. Many argue that the free pursuit of knowledge is being compromised by the need to make it relevant to industrial interests. As Emma Rothschild (1985) suggests, academic institutions are under pressure to set research agendas to achieve this objective: 'the traditional right of faculty to pursue their own research interests is being gradually eroded to maintain the research reputation of the institution'. University scientists are said to be losing their objectivity and credibility as impartial researchers (Epstein, 1979) and greater industrial representation on academic executive boards simply serves this process further: 'The traditional control by the scientific community over most of their fundamental research is being displaced by representatives of private industry' (Ince, 1983, p.59). This is why, in part, we see increasing resources devoted to 'strategic' research.

However, this critique of commercialisation needs to be treated carefully. First, it presupposes that scientists would, were it not for these commercial pressures, pursue research paths in an untrammelled, disinterested and curiosity-driven manner. But, as we saw in Chapter 2, the notion of disinterested research is part of the normative rhetoric of science (see Mulkay, 1976), a selective characterisation of science which acts as a professional ideology serving scientists' interests (see Mulkay, 1979, pp.110–13). Moreover, academic scientists invest their 'scientific capital' (Shapin, 1988) in such a way as to maximise its return for them, which will mean developing research projects which are likely to be tied into the priorities of wider, external funding and policy agencies. As Ziman (1987) has said, 'For most scientists, the problem of "problem choice" is no problem at all: they do research on whatever questions are prescribed by higher authority' (p.95).

Agenda-setting, therefore, is a routine feature of scientific practice and one which is likely to be always influenced by external factors.

Secondly, while it is true that some forms of industrial sponsorship most definitely set the research agendas of scientists – most obviously short-term commissioned or contract research tied to a particular R&D problem an industrial sponsor may have – the most significant feature of the recent pattern of the commercialisation of science has been the development of forms of sponsored research which are very different from such directive contracts. Over the past decade there has been a growing number of research collaborations between academic and industrial partners of a long-term, basic-science nature: these often mix staff from both sectors under the overall direction of an academic director in a broad programme area. They are typically in the bioscience and biotechnology fields, and often related to pharmacological and disease-related research. These hybrid coalitions (Webster and Constable, 1989) involve major (in a few cases, sole) funding of research teams in academic departments by multinational corporations which in return receive exclusive rights over the results of the research. One of the most recent examples is the establishment of a £60 million dermatology research institute at the Massachusetts General Hospital in collaboration with the Japanese company Shiseido.

These collaborations acts as 'discovery groups' for large companies and from the evidence available (Webster and Constable, 1990) it would appear that the coalitions embody a coincidence of research interests between the academic and industrial members rather than the latter dictating the formers' R&D activity. Indeed, it would be contrary to the long-term commercial interests of the companies if they were to be overly directive of research here since this would be to stifle the very basic, discovery-oriented nature of the work.

Third, Etzkovitz (1989), one of the few sociologists to have explored commercialisation in depth, has argued that the process of commercialisation of academic work has not been to its scientific detriment. Instead, he suggests that, rather than being in some way a transgression of the traditional (Mertonian) norms of science, commercial activities herald the *transformation* of norms of scientific practice which allow an accommodation of commercial behaviour without prejudice to the rigour of science: 'through the sorts of structural changes internal and external to the research university . . . a reinterpretation of conduct is taking place such that what had previously been seen as in conflict or incompatible with the proper ways of doing science is seen as in fact compatible' (p.26).

The lesson then is that we should be careful of seeing increased commercialisation as tantamount to straightforward agenda-setting by industry. Nevertheless, it would be foolish to ignore that on a wider front industry has its priorities: any new scientific innovations that impinge on the market holding of companies will need to be either maintained or incorporated (Yoxen, 1986). Perhaps the overall conclusion we should draw is that the relationship between commerce and academia is complex, operating in different ways at different levels of research, within different disciplines of more or less 'maturity', and within different spatial regions, local, national and international. Moreover, 'commerce' is not homogeneous: commercial finance, for example, is likely to have different interests in and place different demands on public sector invention depending on whether it is bank, venture or manufacturing capital.

If science is being increasingly commercialised, it is no surprise to discover an increasing number of sociologists interested in exploring the impact this has had on public sector scientists' experience and conditions of work. Again, Etzkowitz (1989) has analysed the development of new institutional relationships in the United States and their bearing on scientists' work. He singles out the emergence since 1945 of research groups working as *teams*. Team research became important in order to quicken the pace of discovery as competition for priority claims between scientists intensified over the period and as government grants became more focused and short-term. Etzkowitz has argued that these teams which are now the norm in research laboratories are analogous to 'quasi-firms', that is, 'continuously operating entities with corresponding administrative arrangements and directors of serious investigations responsible for obtaining the financial resources needed for the survival of the group' (p.199). He adds that though their origins can be traced back to the professional competition between scientists, they have been a fertile institutional medium within which commercialisation has flourished.

There is still much that sociologists need to understand not only about the impact of commercialisation on scientific labour, but also about the nature of scientific work itself. While there has been a wide-ranging and deep exploration of differing forms of intellectual and manual labour by sociologists, they have not turned their attention to the labour process of the laboratory: as Yearley (1988) has commented, 'The political economy of the modern laboratory has received very little attention' (p. 18).

Finally, it is often said that commercialisation may compromise scientists' access to or publication of research findings. Nelkin (1984) has provided one of the more detailed reviews of this issue, though her examples are almost entirely US-based. Commenting that proprietary disputes are common in biological research sponsored by industry, she draws attention to three ways in which the dissemination of information may be restricted: open communication between what were once informal exchange networks of scientists can become problematic; the peer review process whereby scientists evaluate the contributions made by other scientists to the knowledge base may also be jeopardised if commercial secrecy restricts the release of information; finally, scientists may be less able to publicise findings which may be in the 'public interest' if such findings challenge company practice or products.

It is of course possible that the volume of information does not decline to any significant extent simply because of commercial pressures. Rather, as a commodity information is sold on the marketplace; public sector institutions are themselves selling information or expertise to the private sector and to other public sector organisations. However, what counts as sellable expertise involves both political and technical judgements, so that areas of research not deemed to be so may suffer. This is a genuine concern, but so far inadequately explored within sociology. Moreover, not everyone can afford the price of commodified knowledge, especially in Third World countries. For example, they have been particularly hit by the commodification of gene banks and germ plasm used for plant breeding: stocks of material, once locally and freely available, are now purchased from international agribusiness companies at considerable expense (Kloppenburg, 1989).

Summary and conclusion

In this chapter I have tried to show how the state has developed policies that both support and exploit the science base. A number of key points can be highlighted by way of summary.

1. The exploitation of science varies in different states, reflecting their differing technological bases and political cultures: science policymaking can only be understood through reference to the specific political context in which it is made, contrasts being drawn between contested and closed, pluralistic and planned institutional frameworks within which decisions are made.

2. Distinct political cultures (for example those of the United States and the UK) are likely to lead to different decisions being taken over public interest issues despite similar evidence being available.

3. State support for science and technology has increased steadily in all OECD countries (apart from the UK) over the past decade, and it has been targeted in such a way as to promote an increasing selectivity, concentration, commercialisation and internationalisation of the research effort.

4. As part of the increasingly directive nature of state support for science a range of measures has been introduced to map and evaluate the pattern and productivity of R&D.

5. New external economic and political pressures on scientists mean that they will have to act as 'calculative actors' in the wider policymaking arena, investing their scientific 'capital' in a dynamically changing socio-political environment.

6. The exploitation and support of science and technology is an uneven process, both economically and spatially; regional policies, typically of a corporatist nature, have been developed to iron out the unevenness of technology transfer and economic growth, aiming to revitalise the declining heartlands of capitalism: science parks have been one notable, though often over-glamourised form of regional science policy.

7. Commercialisation and restructuring have had a significant impact on the planning and management of new institutional structures, some of which link academia and industry; while these changes raise questions about the setting of research agendas, the nature and experience of scientific work and the free dissemination of information, the limited sociological research which has explored these issues suggests that the situation is more complex

than depicted in the typical critique that is made of the commercial-isation of public sector research.

One can see from this chapter the way in which science policy geared to the exploitation of research is a complex, multi-dimensional process, shaped by wider political and economic factors. Add to this the fact that policymaking is a contested terrain on which different parties negotiate and construct discourses of 'scientific authority' and 'expertise' as we saw in the previous chapter, then it is perhaps no wonder that the student of the sociology of science policy finds the whole thing somewhat bewildering. Indeed, this is the very nature of the beast as experienced by those in the thick of the policymaking process itself. In part this may explain why 'the best-laid plans of mice and men' (for this is a very patriarchal social arena) often come unstuck, policy appearing disjointed and ineffectual.

This institutional complexity perhaps also explains the attractive-ness of heroic, ideologically fashionable approaches by some governments seeking to overcome the inertia of the policymaking process. In the UK, Thatcherite commitment to enterprise, com-mercialisation and privatisation has been a powerful though glib recipe for institutional change.

Finally, we have seen that the knowledge base of any one country is shaped and exploited according to the specific approach that country adopts towards its support, management and regula-tion. By now, the reader should have abandoned entirely the notion that knowledge is a pool of information in which other scientists and industry fish for new ideas. The more appropriate image might be a marsh or delta, where knowledge of the local terrain is vital, where what lies in the murky waters is uncertain while attempts to channel the flow are expensive and often to the detriment of the local ecology.

5 Exploiting Science and Technology (II)

Introduction

This chapter focuses on the industrial exploitation of science and technology. Although the vast majority of scientists work for private corporations as technicians, engineers or laboratory researchers, the sociology of science has paid much less attention to them over the past two decades than it has to the public sector academic research community. Perhaps in part this is because the earlier Mertonian approach tended to regard scientists in industry as *technologists*, concerned apparently more with applying knowledge for commercial gain than pursuing it via the putative norms of academic science. As such, 'technologists' could be regarded as of secondary sociological importance inasmuch as the key institutional locus for the development of novel, objective science was the academic 'pure' science lab. Indeed, much of the early sociology and history of science sustained this split between science and technology, despite the fact that no one had really explored these two worlds to find out how different they actually were.

Moreover, although it is true that recent sociology of science has developed a much more sophisticated view of both science and technology, as we saw in Chapters 2 and 3, it has failed to engage with the very extensive literature that looks at the relationship between science, technology and the *labour process*. The latter is rooted in a political economy tradition informed by a radical critique of the capitalist technical and social (class) division of labour. There would be much to gain from an exchange between these two research programmes.

Sociologists of science have, somewhat unevenly, explored a number of avenues in their analysis of corporate R&D and the exploitation of new technologies. There are three that have been of particular interest:

(i) a socio-historical analysis of the emergence of R&D within companies;
(ii) an examination of the process of innovation and the social shaping of technological choice, in part filled out through reference to more economistic studies of technological change;
(iii) a less developed exploration of the organisational culture of knowledge within science-based companies.

I shall review this work and end by illustrating corporate exploitation of science and technology through a discussion of biotechnology and genetic engineering. It will be seen that much of this work has its primary focus on the way factors internal to the production process shape and are shaped by science. Little attention has been devoted, for example, to the structure and process characterising the *consumption market* – such as households – for new technological artefacts (but see Cowan, 1987).

The emergence of corporate R&D

Before 1900, the idea of an industrial research laboratory did not exist. Instead, companies bought in expertise to solve technological problems or introduce new ideas. As Dennis (1987) explains: 'Prior to the creation of corporate research laboratories, science and technology were products firms purchased, not processes in which companies invested' (p.482). The principal in-house laboratory was the 'testing lab' where chemists and engineers could determine the quality and precision of raw materials and products against the then-burgeoning national standardisation measures. Today, a large research-intensive company such as the biochemical giant Monsanto will spend over $600 million each year on R&D, or the petrochemical giant Shell $1 billion at its principal lab in Amsterdam, employing over 1700 personnel.

Most major multinationals that depend on an advanced technology base for their survival – and not all do – have development
labs tied to their in-house business divisions as well as one or more
central labs devoted to general and strategic research. The sophistication and capability of these labs not only matches but typically is
far superior to anything found in academic centres, especially in
terms of standards of equipment. What caused these changes in the
corporate R&D system over the past century?

The past decade has seen a steady growth in the number of
studies that have examined the growth of the research within
companies. Some studies have provided detailed analyses of individual corporations, such as Hounshell and Smith's (1988) account
of Du Pont, the world's largest chemical company, Reich's (1985)
of General Electric, or Swann's (1988) on the American pharmaceutical industry, particularly R&D strategy within Eli Lilly. There
have also been more general synoptic reviews of this history, most
recently by Mowery and Rosenberg (1989). Despite their differing
substantive focus, they have a number of points in common, as
they trace the companies' organisational structures and strategies.

First, the move towards in-house R&D was prompted by the
need to rationalise and target research lab activities on those areas
on which the future competitive success of the corporation
depended. This was especially true for firms in the chemical and
electrical industries, where growing competitive innovation had
been accompanied by a rapid expansion of patenting activity.
Rather than buying-out patents held by others or developing new
products through licensing, many firms sought to develop their
own patentable inventions: Reich (1985) quotes the words of GE's
chief patent officer: 'If someone gets ahead of us on this development [i.e. a new mercury vapour lamp] we will have to spend large
sums in buying patents or patent rights, whereas if we do the work
ourselves this necessity will be avoided' (p.66).

Secondly, the companies were financially strong enough to
entertain the idea of new general research labs: they all had large
markets for their products, despite the competition. This enabled
them to take a long-term view towards their innovation strategy,
prepared to invest in R&D that might not have a pay-off in terms of
successful products for a number of years. As today, small
research-intensive companies are thin on the ground, tending to be
limited to the very small high-tech enterprises such as one finds on

some science parks, feeding their ideas into or perhaps being bought up by larger corporations.

Thirdly, the turn of the century saw a rapid growth in the number of those with a formal higher education in the new sciences, especially in Germany and the United States. The German higher education system was decentralised and competitive, providing new opportunities for an expansion of teaching and research, the development of new specialties and the emergence more generally of the professional scientist. The United States emulated the German approach. Corporations had, therefore, a new supply of trained researchers and technologists who could be employed in the new labs.

Finally, as Mowery and Rosenberg stress, the period saw an extensive restructuring of capitalist enterprise with growing internal differentiation of the business into new product and support divisions, including research: as they say, 'the expansion of industrial research was linked as both cause and effect with the reorganisation of the American corporation during the late nineteenth and early twentieth centuries. Technically trained managers, a strong central staff able to focus on strategic, rather than operating, decisions such as marketing – all were associated with the growth of R&D within the firm'.

Freeman (1982) has classified corporate R&D into two types: defensive and offensive. The first refers to an R&D strategy that defends a company's existing technological capabilities and advantages over competitors, perhaps by continual marginal improvements over its products and techniques, perhaps by incorporating innovative technology from other firms that poses a threat to it. An 'offensive' research on the other hand, means that companies are prepared to support risky, unpredictable knowledge-oriented research in pursuit of new scientific and technological development. In fact, the larger companies are likely to pursue both strategies located in their development and basic research labs respectively. The Du Pont study illustrates this dualistic approach, one that appears to have served the company well in terms of staying ahead of its competitors in the development of new synthetic fibres, including neoprene and nylon. As research programmes mature, however, and new products become increasingly difficult to create, 'offensive' R&D can be perceived as increasingly expensive by senior corporate managers. This happened for a time during the

1960s at Du Pont and greater attention was then given to the development and better marketing of its commodities. The relationship between defensive and offensive R&D strategies is then complex and dynamic. Moreover, simply because corporations are prepared to invest in state-of-the-art strategic research, such as genetic engineering, we should not assume that they do so in order to initiate sweeping changes in their own, and the more general, technological base. As I shall discuss more fully when I examine biotechnology, it is possible to argue that multinationals' investment in this area has been more 'defensively' than 'offensively' inspired.

However one regards the strategic *purpose* of research, its *character* within large companies is a mix of both basic and applied science: in fact, as we saw in Chapter 1, this very distinction is breaking down as strategies for technological development in the new sciences stress interdisciplinary skills. But the more companies adopt this approach the more difficult will it become to sustain, as many do, a product division structure underpinned by discrete technological bases. At the end of the twenty-first century, those writing about the organisational structure of corporate R&D may well report the emergence of a very different pattern for managing it.

Given the increasing size and capital value of multinational corporations, they continually monitor the efficacy of their in-house management of R&D. During 1988–90 there was a spate of takeovers and mergers, especially in the pharmaceutical sector as multinationals responded to increasing costs of R&D, threats to their market through the development of generic drugs and downwards pressure on pricing policies in the US and Europe. This restructuring will eventually work its way down to the research divisions and involve further rationalisation of the research process. It is increasingly likely that companies will be forced to emulate the well-established practice in Japan of collaboration between large and small companies in the new technology areas. But it is unlikely that the real value of R&D will decline, especially in the science-based industries, such as chemicals and electronics, although this varies by country. Table 5.1 shows R&D expenditure for some key sectors in the UK.

These bald statistics, although telling of a general growth in R&D, say nothing of the specific directions that this research has

Table 5.1 *In-house expenditure on R&D in broad groups of industry (£m at 1985 prices)*

	1981	1983	1985	1986	1987
All manufacturing	4393	4208	4673	4895	4994
Chemicals	772	812	941	1002	1199
Electronics	1545	1630	1758	1882	1707
Motor vehicles	225	264	371	380	414

Source: *British Business*, Department of Trade and Industry, Feb. 1989

taken, nor the circumstances surrounding the decisions to go one way rather than another. Technological strategies are chosen to reflect the specific priorities of companies: there is nothing inevitable or predetermined about the way technological systems within companies and economies as a whole are established, as Hughes (1987) has shown at some length. Although they may be large-scale, consolidated, supported by a whole range of practitioners, bureaucrats and consumers, they are never autonomous from the continual social shaping that gives them their stability, or (as Hughes calls it) their 'momentum'. It is this idea of the social construction and maintenance of technology that underpins the second main theme I want to explore briefly here, that is, the sociological examination of innovation.

Innovation and technological choice

In classical (and contempory neo-classical) economic theory, the relationship between technology and industrial growth has been understood in terms of the so-called 'production function'. In brief, the theory assumes that given a certain level of technology available there will be a more or less efficient ratio of capital to labour in the production process, and hence more or less efficient levels of productivity. It is assumed that the adoption of new techniques in production leads to economic growth. In this context, R&D is obviously regarded as an important part of the production function.

However, as Clark (1985) notes, a number of economic analyses that have sought to measure the statistical contribution of R&D to growth have been unable to find this simple relationship between the two assumed by the traditional theory. Other factors, such as levels of education or economies of scale, seem at least as important. But more generally it has been increasingly recognised by analysts of technological change that the process of innovation and its relationship to development is a complex matter, that *innovation is a process that involves continuous feedback* between its constituent parts. Thus, it is probably mistaken to seek to identify *discrete elements of the innovation process (such as R&D)* in order to determine their respective importance for economic growth.

What then, is the importance of R&D? And why all the attention given to it in today's science policy programmes? Clark answers this through analogy: 'R&D has the properties of an escalator moving downwards at an increasing rate through time. The firm may have to climb it since otherwise it goes out of business, but the effort required actually to move up becomes progressively greater' (p.127). Thus, the policy implications become clear: 'In many sectors firms *must* invest a given absolute sum in R&D, the precise amount determined by the nature of the industry in question, simply because if they do not they will find it increasingly difficult to compete with their rivals. Investment in science does not so much 'push' innovation as act as a necessary condition for industrial survival' (p.140).

The motivation for engaging in R&D, its strategic role in the corporation is, then, in the 'real world' less straightforward than the classical theory tends to suggest. Moreover, technology analysts have placed more stress recently on the need to relate social, political and organisational factors together in any model of technological change. Freeman is one analyst who has advocated this more synthetic approach: 'Although some enthusiasts continue to advocate the use of the production function approach, most economists now seem more sceptical about the feasibility of this method, and increasingly, about the theoretical assumptions underlying the work. Studies based on this method may be particularly criticised for their failure to recognise the importance of complementarities in social and technical change, for their neglect of all other social science disciplines, for their lack of historical sense, and for their reductionism in relation to "technical change" ' (1977, p.244).

Freeman, and more recently Coombs *et al.* (1987), are critical of
the traditional neo-classical position, which tends to 'black box'
technology, that presumes there is, as Clark has said, some
'technological shelf' available from which businesses can choose
that technology most appropriate to their needs. We saw in
Chapter 2 how sociologists, working within the relatively recent
fields of 'the social construction of technology', have challenged the
black-box image of science and technology, (Latour's 'technoscience'). Technology is continually being shaped by social processes
reflecting the different perceptions and interests of individuals and
social groups. This shaping is not a random affair: in part, it is
dependent on the *power* that different groups involved bring to the
innovation process.

The relationship between power, choice of technology in production and the wider work relations that follow has been considered much more fully by sociologists, even though this analysis
has had only limited impact on the sociology of science itself. This
work explores the way capitalist (in contrast to non-capitalist)
modes of production apply scientific knowledge and its associated
technological changes to the process of production. Typically, this
has meant a gradual increase in productivity and growth
accompanied by increasing control over the skills and labour of the
worker. While this process can be traced back to the reorganisation
of industrial production during the first industrial revolution, the
analysis locates the first real step towards an effective and clinical
control of worker-power in the early twentieth century with the
appearance of Frederick Taylor's principles of 'scientific management'. Taylorism, as it came to be known, developed precise
measures of the time and motion needed to perform any particular
industrial task. To maximise productivity, every detail of the
workers' jobs was specified, as well as 'necessary' periods of rest,
the time it might take to go from one part of the job to another, and
so on.

While Taylor's ideas spanned the first few decades of this
century, many radical analysts argue that their impact is still felt
today, not only in heavy manufacturing, but also in the lighter
industries and white-collar office work. Braverman (1974), offers a
most forceful Marxist critique of contemporary Taylorist principles, arguing that the whole basis of capitalism required the
continual 'deskilling' of the workforce. Workers' knowledge and
skill would be squeezed out of them by either breaking their

work-task up into smaller and more easily-performed components or using machinery to mimic the work they do. Since Braverman's text appeared, there have been many studies by sociologists designed to assess the degree to which modern capitalist production *necessarily* requires the deskilling of labour. Although there is evidence showing that the process does occur, it is much less clear that it will always, and indeed there is some evidence showing that in certain circumstances it may be counterproductive: that is, sometimes capitalist employers will seek to control their workers by ensuring that they feel a strong sense of identity with the company, or by recognising a special class of workers' skills which will receive high reward over and against a broad mass of employees who can be recruited and dismissed easily, who will be paid much less, and who are likely to have little sense of loyalty to the company: this sets up a two-fold or *dual* labour market, a high-skill, high-pay, stable sector, against a low-skill, low-pay, unstable one.

These modifications of Braverman's thesis are, however, only modifications: they do not seek to challenge the broad thrust of Braverman's critique of modern capitalist technology. Noble's (1985) study of the introduction of 'numerically controlled machine tools' in the metal working industry during the 1950s provides a good illustration of the more subtle, post-Braverman analysis. The study illustrates the way in which a particular machine was developed and then incorporated into the production process as part of a wider set of strategies being deployed by management to increase their control over it. Prior to the introduction of 'numerical control' (N/C), machine tools, devices used to cut and shape metal to be used as tools themselves or as parts of other machinery, had to be operated by a skilled worker who would control the cutting pattern according to a predetermined design. To increase the pace of machining, either more workers would have to be employed or the special skills of the machinist would have to be automated. The latter first became possible through N/C which worked as follows: 'The specifications for a part – the information contained in an engineering blueprint – are first broken down into a mathematical representation of the part, then into a mathematical description of the desired path of the cutting tool along up to five axes, and finally into hundreds or thousands of discrete instructions, translated for economy into a numerical code, which is read and translated into electrical signals for the machine controls. The N/C tape, in short, is a means of formally circumventing the role of the machinist as the source of the intelligence production' (1985, p.111). N/C allowed management to secure the control of the production process.

However, unlike Braverman, Noble records that managerial control is never complete, even when automative technology like the N/C has been introduced. Management, in this case as in many others, found they needed experienced and skilled workers to oversee the machines because they were liable to break down and contribute to a highly expensive machine-driven smash-up. Thus, management's initial strategy to deskill and displace specialist machinists was quickly revised to ensure proper supervision of the new equipment. As Noble comments, the workers' intelligence, knowhow and experience is not something that is best removed from the shopfloor. On the contrary, private capital needs to draw on the initiative and skill of its workers in order to remain productive: as Cressey and MacInnes (1980) emphasise, capitalist production cannot rely simply on the coercion of labour, it must at the same time try to incorporate and reward worker skill. Indeed, technological change may require the development of new technical skills and not merely the incorporation or enhancement of older ones.

Clark *et al.* (1988) show, for example, that the introduction of a new telecommunications exchange in Britain during the 1980s, the TXE-4 semi-electronic switching system, meant that British Telecom maintenance engineers and senior technicians found their work tasks and skill-inputs increasingly demanding compared with the pre-electronic electro-mechanic exchange system. With the electronic technology it became much more difficult to locate faults, much more necessary to understand the system as a whole; once found, however, faults were technically simpler to repair. The authors report that management (senior supervisors) found their authority weakened as they had less training in and understanding of the new technology. This suggests that certain technological choices might have unintended effects inasmuch as they reduce the managerial power of those in senior positions: that is, sometimes those in authority may find unexpectedly that a new technology cuts across the ways in which they have traditionally organised and directed work.

Nevertheless, all these studies show how the priority for corporate management is to ensure a compliant yet educated and educable workforce whose own knowledge of the production process can be drawn on to keep the enterprise competitive. The point is, this does not stop at the production end, on the shopfloor, but can be traced upstream to the R&D labs of the same corporations. Within scientific research labs, personnel-management tension and conflict over the pace and direction of research can and does appear. White-collar, 'intellectual' (as compared with manual) labour will experience in their way the contested terrain of industrial relations. And just as automation and robotics characterise a large number of production facilities so they are appearing in R&D labs of the bigger companies: 'At the Upjohn Company's laboratories in Kalamazoo, Michigan, an ambitious system called Generic Analytical Sample Preparation (GASP), is being developed to handle one or as many as 100 samples at a time. To accomplish this the system follows a flexible, factory-line approach to robotics . . .' (Newman, 1990). In this context it is the technicians who are most likely to be 'deskilled' by automation, and junior scientists who constitute a large pool of short-term flexible labour: unfortunately, unlike office and factory workers, sociologists have conducted very few detailed studies on this class of workers. Much has been written at a general level (e.g. Sohn-Rethel, 1978; Palloix, 1976) along lines similar to Hales' (1986) comment that: 'A large proportion of the scientists employed in modern firms are heirs to [the Taylorist] tradition. As skilled mental workers, it is their job to ensure the evolution of technical knowledge and its embodiment in material form as technology, this responsibility having been removed from the unreliable "hands" of the traditional workforce and entrusted to the controlled hand of scientific research' (p.79).

In order to sustain these arguments, sociologists need to explore much more fully the work relations that constitute the political economy of the R&D division of private sector companies. We would then be able to see how the exploitation of knowledge varies from one part of the production process to another, whether, for example, the routinisation and fragmentation of scientists' labour is much less than downstream in the factory. In short, we need to know more about the *culture* of work relations within corporate (and not just academic) labs. There has already been a start in this direction, as the next section describes.

The organisational culture of knowledge in corporate R&D

The rationalisation, standardisation and automation of work-tasks that one finds in manufacturing divisions of companies are much more difficult to achieve in the setting of the research lab. They are present, though, to some extent: one finds most labs will use technicians or junior scientists to do very restricted, monotonous and repetitive tasks all day every day, whether it be cloning genes, purifying enzymes or whatever, without any real involvement in the research process itself that makes use of these materials or techniques. However, the pressure to introduce rationalised, bureaucratic forms of control over scientists' labour will tend to undermine what that labour is supposed to produce: novelty and commercial discoveries.

There have been very few sociological studies of the organisational culture of the R&D lab. What there are focus on two different aspects of the culture: the strategies that research *managers* adopt to encourage the greatest productivity from their research scientists; and secondly, the strategies research *scientists* will adopt to establish the conditions under which they can engage in intellectual interesting research while simultaneously meeting the company's demand for commercially relevant research. It is often the case that commentators on the commercialisation of science regard these two objectives as mutually exclusive. However, this is not necessarily the case, as has been suggested by Bartels and Johnston (1984) who note that there can often be a 'proximity' between disciplinary and commercial objectives such that both their needs are satisfied. Let us look at the first group of these studies, those that focus on the management of R&D.

One of the central debates in this literature concerns the way in which management might try to organise R&D. There are two contrasting approaches that management might take which sociologists have explored: these are the 'mechanistic-bureaucratic' as opposed to the 'organic-professional' approaches. This distinction was first elaborated by Burns and Stalker (1961). In the former, R&D is managed in a highly rigid, hierarchical structure embodying specialised work-tasks answerable to a central authority (the research manager). While this might be appropriate for the shop-floor, it is too mechanical or mechanistic a structure to impose on scientists since it creates a cumbersome and inflexible R&D culture that limits the pace of innovation and the response to innovations elsewhere.

The second, 'organic-professional' approach is the exact opposite of a mechanistic style: it encourages much greater horizontal and vertical integration of scientists (on a face-to-face basis) whatever their formal authority in the organisation, and a much faster response to both new opportunities as well as threats from competitor corporate labs. Similar distinctions about the form of management control have been made by other analysts, such as 'programmed' and 'non-programmed', or 'authoritative' and 'participative' types of management. The problem, of course, from an organisational viewpoint, is sustaining the organic approach while corporations grow ever larger through the accumulation and concentration of capital. Recognising this, many but not all companies have tried to maintain their innovation rate by separating central R&D labs from the more manufacture-oriented development and production labs and business divisions. This can create as many problems as it solves as it deliberately breaks the relationship between strategic research and the immediate concerns of commercial development and productivity. Intermediate mechanisms have then to be put in place to provide the institutional linkage between the two in order that innovation upstream can be converted into marketable products or processes downstream.

At the same time, organically managed R&D is not necessarily always better than the more mechanistic counterpart: much depends on the state of the particular sector or market towards which it is directed. As Hull (1988) has said: 'Organisation [of R&D] should match environmental requirements. Inflexible forms, such as the mechanisatic type, fail to survive in dynamic, high technology markets. But adaptive forms, such as the organic type, also fail to survive in maturing markets unless they become more structured as they grow' (p.395). Hull has provided a summary of the important distinctions between the two approaches, summarised in Table 5.2.

Hull tests which of these R&D structures is more conducive to innovation by measuring the patenting activity of 110 medium-to-large-size companies in New Jersey (USA). He concludes that the best innovative context is one that combines both mechanistic and organic approaches. He argues that this is a characteristic feature of

Table 5.2 *Organizational systems: mechanistic (M) versus organic (O)*

Type	Environment	Work inputs	Throughput technology	Organisation structure	Perfomance outputs
M	Stable, homogeneous, mature products	large-scale non-complex, e.g. large quantity of similar items for assembly	mass production, e.g. sequential work flow	mechanistic–bureaucratic	low innovation, high productivity, e.g. metal beverage cans
O	dynamic heterogeneous, new products	small-scale, complex, e.g. R&D intensive problem involving unknown qualities	complex batches, e.g. reciprocal feedback work flow	organic–professional	high innovation, low productivity, e.g. unique aerospace device

Source: F. Hull (1988)

many Japanese companies and has allowed them to be more innovative and productive compared with corporations elsewhere. Organic organisation is most likely to be effective in generating innovation in small-scale operations (perhaps the typical science park company). Clearly, the relative balance between the two approaches is not then, simply a matter of an employer's personal preference for one style of management rather than another. The strategies chosen for exploiting scientific knowledge within the company reflect the organisational and wider environmental contexts within which the corporation has to operate.

This institutional need for a balance between mechanistic-bureaucratic and organic-professional opens up, however, the possibility of greater room for manoeuvre for R&D staff insofar as management struggles to make this balance work. How scientists respond to both the managerial constraints on them as well as this space afforded to them for professional autonomy is in need of much further investigation by sociologists. In particular, the social constructionist perspective needs to examine the experience of constructing science and 'exploitable' technology in the *applied corporate context*. We would then be in a better position to explore scientists as calculative actors responding to the organisational and technological demands of industry.

One study that has made some attempt to do this is that by Fujimura (1987). She argues that scientists will always tackle those research problems which are more likely to be resolved, or are more 'do-able'. But simply having the technical skills to complete the experimental task successfully is not enough. Researchers have to engage in a whole range of associated tasks connected with the management and presentation of their work to others within and outside their laboratories – to R&D managers, sponsors, other scientists and so on – and only when all these other important levels of the scientists' 'social world' are successfully negotiated can scientists regard their work as being 'done': 'scientific work gets done and problems are solved when all the necessary parts at all levels of work organisation are collected and made to fit together'. Doing scientific research is then like completing a three-dimensional social jigsaw aligning the different demands and tasks that the scientists confronts at the experimental, the wider laboratory and finally the social world (which includes the wider field and type of organisation in which the experimental work is set) levels (see Figure 5.2). When 'things don't work out' it is because all three levels are unsuccessfully aligned. The analysis here reminds us of Latour's (1987) account of scientists negotiating in various competitive arenas from the lab bench onwards the status of their knowledge claims. As Fujimura herself says, '[T]his notion of alignment is similar to Latour's "enrollment of allies" and "keeping [these allies] in line" in the construction and maintenance of scientific facts' (p.264).

Scientists working in corporate labs have to satisfy different demands from different 'social worlds', for example, the 'world' of the employer or sponsor and the 'world' of the broader scientific discipline of which they are members, say, molecular biology, biochemistry engineering, and so on. Meeting both types of demand requires some manoeuvring: in her own case study of a corporate biotechnology lab Fujimura reports: 'When the demands from the two social worlds conflicted – that is, a product deadline versus standards of good science – lab members tried to devise strategies which met both worlds' demands "well enough". Their main strategy was to divide up the labour of product development and pass some tasks on to other departments. This allowed them to continue their biochemical tasks' (p.275). In general, therefore, scientists sought to accommodate sponsors'/employers' own commercial demands within their specific intellectually-driven research interests.

Most importantly, Fujimura stresses that it would be wrong to regard sponsors' interest as somehow clear and fixed: they too enter the negotiating arena of the three levels she identifies and in the process become reshaped and perhaps redirected, in part as a result of the efforts of research scientists themselves:

[S]ponsors' goals are not necessarily hard and fast, especially when they are mediated by several agents. In this case, the scientists negotiated and even bartered to get their project accepted. [One scientist is reported as saying]
We have this thing called the [committee name] where people with new ideas stand up and chuck them out. [The ideas] get sort of thrashed around . . . If people like it, then there's an assessment from marketing people as to what the payoff will be and from Research people [whether it is] feasible. And then some sort of bartering goes on and so on. And people decide

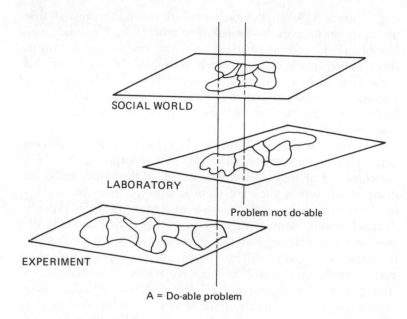

SOCIAL WORLD

LABORATORY

Problem not do-able

EXPERIMENT

A = Do-able problem

View from above

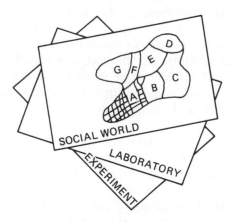

Fig. 5.1 *A metaphor for aligning tasks at three levels of work organisation*
Source: Fujimura (1987)

roughly what sort of support it should get in terms of people and the layout of money, based on those facts. [Can marketing say no?] No, they would say there's no money in it. Marketing has looked at this [oncogene] project and said, 'We don't see any chance of anything we can market.' But if Research says, 'Okay, fine, but we think it's good for our image to do this research, and to hell with Marketing', those things will be taken into account. So Marketing is . . . only one element. The visibility, getting publications out, developing technology – there [are] a lot of other justifications for doing a project (p.266).

Fujimura's study provides one of the few detailed illustrations of how *within corporate* labs the process of the negotiation of research agendas prevails just as it has more often been shown to do in academic contexts. At the same time, she acknowledges that it is more difficult for scientists in industry to adjust to a situation in which sponsors ultimately decide to move out of an area of research altogether: 'Larger economic changes, for instance, may result in loss of funding for a laboratory, a line of research or a whole discipline. In these cases, researchers may choose to readjust their work . . . [although] such readjustments are not trivial' (p.280). This is surely something of an understatement where, for example, scientists find corporate restructuring means the complete abandonment of a whole research field: looking for a new job is unlikely to be experienced simply as a 'readjustment'. Nevertheless, in conditions of more secure employment, Fujimura's account indicates that corporate exploitation of knowledge upstream in the laboratory is a complex, interactive affair. Compared with the situation confronting technicians and shopfloor workers in the production divisions downstream, however, research teams have more opportunity and perhaps power to negotiate their terms of work on the particular 'contested terrain' they occupy.

This is confirmed in another recent study by Vergragt (1988) of two industrial research laboratories in the chemical industry. He shows how the actors involved – the senior scientists, marketing personnel, economic advisers, and so on – adopted various strategies to safeguard their professional interests and secure the particular 'niches' they occupied in the corporate culture of the companies. They used their specialised technical skills as forms of leverage or power to direct technological innovation down paths which they regard as most appropriate. Vergragt examines the way decisions are taken as a result of this process. Decisions about research paths are typically made in response to 'critical events' – such as the introduction of new state regulations controlling the field of research – or the more focused 'critical *research* event' such as the outcome of an important set of experiments. He emphasises that the social actors involved will try to ensure that all such events are perceived and interpreted in ways which suit their particular occupational interests in the overall corporate enterprise. Obviously some are 'winners', some losers, because of differences in power and resources they draw on in the negotiation process. Like Fujimura's model that depicts the way scientists align differing worlds through a process of articulation and negotiation, Vergragt's own model highlights the calculation, negotiation and decision-making dynamics of the corporate research process (Fig. 5.2).

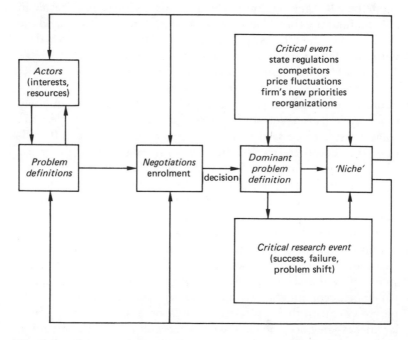

Fig 5.2   *Conceptual framework of the industrial R&D process*
Source: P. Vergragt (1988)

So far in this chapter, I have discussed the contribution that sociologists have made towards our understanding of the origins, form and dynamics of corporate exploitation of scientific R&D. As we have noted, much more needs to be done in this area while considerable benefit would be derived from more exchange between the sociology of science and those sociologists working in the sociology of work tradition. There is however, one area of corporate R&D which has received considerable attention from sociologists (and many others). It is a field which is particularly interesting inasmuch as it has raised questions about the ethics of research, policies towards corporate and academic linkage, the appearance of a new growth cycle in capitalism, policies towards regulation, and so on. This is biotechnology.

Biotechnology: setting new technological and social agendas

Biotechnology is often hailed as the new 'revolutionary' science. Undoubtedly, as I shall describe shortly, it has meant some very big changes in certain fields of strategic and applied science as well as ushering in new high-tech biological products. Like any other technology, however, its impact and direction reflect the interests of those making use of it and not some asocial unfolding of an 'inevitable' scientific path. Revolutions depend on the interaction of human agency, and social structures and technological revolutions are no exception. There is nothing *necessarily* 'revolutionary' about any technology: it depends on the way it is adopted by people in specific social circumstances.

Moreover, the apparent 'novelty' of biotechnology in the late 1970s can also be overplayed as, like any other field, it has depended on prior developments in biology and biochemistry. At the same time, however, simply because this is so we would be wrong to assume, for example, that all biologists in the late 1960s would have considered genetic engineering, the most significant biotechnology, a serious proposition, a 'do-able' problem. Some, such as Stent (1968), even believed the heyday of molecular biology to have come and gone. In short: beware all linear histories of science.

Before we can examine the debate over biotechnology we need to know what it is. There have been various attempts to define it. A recent one from the European Federation of Biotechnology (1988) will do here:

> The integrated use of natural sciences (e.g. biology, chemistry, physics) and engineering sciences (e.g. electronics) by the application of biosystems (cells of microbial, plant and animal origin) in bio-industries in order to supply biosociety with desirable products and services (p.207).

The definition indicates that the field is strongly interdisciplinary and industrially oriented. Both characteristics have informed the development of one of the most important biotech fields, molecular biology, since its emergence in the 1930s. Its growth has depended most particularly on a combination of biological and chemical engineering principles, the former providing an understanding of the basic processes underlying organisms, cells and molecules, the latter a mathematically-based control technology which helps translate the basic work of the geneticist into large-scale industrial products and processes. There are a number of specialty areas where 'the biotechnologist' might be found working. These include:

- Recombinant DNA genetic engineering (particularly associated with plant, animal, and drug development).
- Enzyme and biocatalyst research (important in areas such as food processing, biosensors and diagnostic kits).
- Process engineering (related to research on recycling, recovery, waste extraction, effluents and so on).
- Cell culture and single cell protein production (relevant for biomass production and the development of fine chemicals [such as steroids]).

This list is by no means exhaustive, but does give an indication of biotechnology's potential scientific and industrial impact. I shall focus my attention on the first of these because it has often been regarded as the cornerstone of the biotechnological 'revolution'.

In order to recombine or engineer genes one needs to understand the structure or architecture of the cell and the molecules that constitute it. As with all other areas of science, there has been considerable dispute and debate about this, continuing through to the present day (see, for example, Hillman, 1972). Nevertheless, there is a dominant paradigm of the cell which is reproduced here (Fig. 5.3).

Genetic engineering involves the manipulation of genes that are located within the chromosomes in the cell nucleus. Virtually all genes are made of the substance *deoxyribonucleic acid (DNA)*. The existence of DNA and its role in genes had been discovered in the 1940s, though a consensus over its structure emerged only with the publication of Watson and Crick's 1953 double helix model: they proposed that the DNA molecule is made up of two strands wound around each other. Along each strand are located four base chemicals, adenine (A), thymine (T), cytosine (C) and guanine (G). While the order of these chemicals along one strand may vary as you move along it, each is tied to a complementary chemical on the other strand such that A on one strand only bonds to T and G only to C (see Figure 5.4). These form the 'base pairs' of the DNA. If a strand is unwound it can only bind with another strand with a complementary run of the four base chemicals: this means that

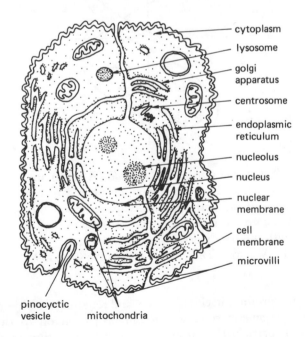

Fig. 5.3   *The cell*

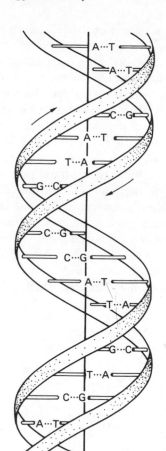

Fig. 5.4 *Watson and Crick's DNA model.* A schematic illustration of the double-helix. The two sugar-phosphate backbones twist about on the outside with the flat hydrogen-bonded base pairs forming the core. Seen this way, the structure resembles a spiral staircase with the base pairs forming the steps.
Source: J. D. Watson, *The Double Helix* (1968), p.159

DNA can reproduce itself accurately and repeatedly within the cell providing a genetic code which can be passed on to the next generation, particularly through the enzymatic action of protein molecules which are manufactured through the instructions of

DNA. While DNA provides the basic instruction or template for how an organism's genes will look, proteins perform the specialised biochemical tasks within an organism. Each protein has a specific structure built from amino acids which will determine how it will function. The relationship between the structure and function of proteins is of major interest to biotechnologist because, once these are understood, it would then be possible to engineer proteins as completely new biological catalysts.

DNA is both a very complex but very thin molecule which allows it to hold so much information in the small chromosome. As Fairtclough (1985) shows, even a 'simple' molecule like the bacterium *Escherichia coli*, is architecturally complex:

'The fairly simple bacterium *Escherichia coli* has its genetic information in a single chromosome which is one long double helix of DNA. It has four million base pairs . . . The total length of the *E.coli* chromosome is 1.4 mm, but it is wound up in such a way that it fits into the bacterial cell only a thousand of the length. This would not be possible if DNA were not a very thin molecule. In fact, in the case of *E. coli* the helix is seven million times as long as it is wide. At this ratio of width and length an ordinary piece of thread would have a length of over half a mile. In more complicated organisms there is a lot more DNA. In each human cell there are six billion base pairs of DNA and a total length of DNA of about two metres. The actual DNA molecule is so thin that one big enough to reach from the Earth to the Sun would weigh only half a gram!' (p.60).

Given the scale and the complexity of the biochemistry of DNA, the idea of unwinding it, isolating specific fragments and coupling them to other pieces of DNA which carry different genetic properties in order to confer a desired characteristic on an organism might seem somewhat ambitious. It is certainly exceptionally difficult (though not impossible) in some areas such as plant and animal species: one must be able to insert the desirable gene into the host organism and to ensure that it will be properly 'expressed' or function correctly in that organism. The latter is, of course, crucial for there is little point in introducing new genetic information (say related to disease resistance in plants) if the host organism fails to turn it on.

This is one reason why biotechnology has had a much greater impact on corporate R&D within the health sector where genetic engineering of synthetic drugs is technically, ethically and commercially a less risky investment. Even here, the impact of biotechnology on the pharmaceutical industry has been limited and has certainly been much less than the hyped-up forecasts made during the mid-1980s. There are currently only about a dozen prescription drugs available which have resulted from biotechnology, accounting for less than 1 per cent of global pharmaceutical sales.

I shall now move away from the specific science of biotechnology to the sociological issues it has raised. (For those wishing to pursue the investigation further, there are a number of introductory texts which are user-friendly to the non-specialist like me (see e.g. Yoxen, 1983; Primrose, 1988, Trevan *et al.*, 1988).

Biotechnology might be said to involve three changes which are often regarded as 'revolutionary': one in science, another in industry and a third in the relationship between science and the wider society. It has, in other words, set new technical and social agendas, reflecting, of course, the different interests of those who develop, commercialise and seek to regulate it.

As a revolution in science, it can be said to have had two effects: first, the breaking-down of biological barriers, and second, the breaking-down of disciplinary barriers. Biologically, the advent of genetic engineering has brought about the possibility of controlling genetic information such that, for example, one can transfer genes between species that in nature are intersterile, or manipulate and remove mutant genes that might be associated with genetic diseases, or construct novel proteins that have never existed naturally. As a field, biotechnology has demanded an interdisciplinary mixing of specialties in a way and to a degree never seen before. Moreover, as was hinted at in Chapter 1, this has meant that the distinction between basic and applied research no longer seems appropriate. As Panem (1984) has pointed out, biotechnology has created the phenomenon of the 'time collapse' whereby fundamental research becomes immediately commercially relevant.

These two changes are interrelated inasmuch as breaking through the biological barrier required transdisciplinary expertise. We are then talking about important changes in the ways of doing and perceiving science: while these shifts are felt most strongly in the epicentres of the biotechnology labs of research institutes and private companies, the institutional shock waves may well be registered further afield as disciplinary-based science is confronted by an increasing number of cross-cutting patterns of research that build new networks or 'core sets' of scientists (see Chapter 2). To the extent that this process does occur, it will gain added momentum from the commercialisation of public sector research (see Chapter 4) which has been especially significant in the biotechnology field. As we saw, corporations are interested in establishing new types of institutional relationships with academia. It is conceivable that with further state support, such relationships could become generalised beyond the immediate domain of biotechnological research.

What of the notion that biotechnology is the basis for a new 'industrial revolution'? Many have argued that we are witnessing the appearance of a whole new type of economy which will be increasingly based on the innovations of biotechnology, whether in food production, health care, energy provision or waste treatment. Some have labelled this the new 'biosociety' (European Commission, 1983) which will not emerge as a gradual evolution from the existing economic and technological order but through a radical, innovative break with the past. As such, biotechnology has been called a 'locomotive technology' (Sharp, 1986); that is, one that combines a series of major innovations which together initiate a fundamental change in the technological infrastructure of society.

However, without doubting that biotechnology has had and will continue to have a significant impact on society, the idea that it heralds the arrival of an entirely new socio-economic system can be questioned. Measured simply in terms of the capital investment and return in the area, biotechnology has not had as spectacular a growth as some commentators predicted. Forecasts made in the early 1980s have been continually revised downwards in real terms as the problems of translating invention into commercial innovation have built up: in one or two specialist areas, the failure to realise Panem's 'time collapse' has resulted in a corporate collapse. Commercial relevance is a necessary but not sufficient condition for commercial success.

Secondly, Buttel (1988) has argued that biotechnology fails to meet the criteria for a revolutionary technology. For innovation to be revolutionary it needs:

– wide applicability in an extensive range of product areas
– to be incorporated extensively in the manufacturing process to reduce production costs

- to create entirely new consumer goods
- to link with emergent industrial sectors which have sufficient financial capital available to invest heavily in the new technology.

Buttel believes that the electric motor in the early decades of this century met all these criteria and among other things, 'permitted mass, assembly-line manufacturing, made possible the formation of the industrial working class, and led to whole categories of new products (especially consumer durables such as refrigerators, vacuum cleaners, washing machines, and so on)' (p.9). He suggests that microelectronics and information technologies will have a similarly transformative effect. But biotechnology, he says, will not be a locomotive technology but one that is 'subordinate' and 'subsidiary', used largely in defensive ways by companies, as a 'substitute' (rather than revolutionary) technology in an effort to maintain their competitiveness in increasingly tight or even declining markets. Thus, Buttel claims that 'the development and application of biotechnology will be primarily to cheapen or otherwise improve the production of *existing* products and/or provide substitutes for existing products or services' (p.6). In agriculture, for example, Buttel argues that the improvements brought so far have been merely to serve to 'patch up' the problems of Western agriculture and the related petrochemical (for fertiliser/pesticides, etc.) industry. He accepts that its role in the health (and drug) sectors will be important but argues that relatively few biotechnology-derived commercial products have been developed, and even where they have, their long-term future depends heavily upon the maintenance of high drug prices. As is increasingly acknowledged by the pharmaceutical industry itself, this is looking less certain as the switch towards generic from proprietary (brand name) drugs gathers pace in the big markets of the USA and Europe.

Buttel is not denying that biotechnology will have a significant impact on economies, simply that this will be a gradual, evolutionary rather than revolutionary one precisely because of the way it is incorporated into existing social and economic systems. Revolutionary technologies are those that are caught up in and integral to wider forces of social change which are themselves radically transformative.

If one looks at the historical circumstances in which companies began to invest large amounts of capital in biotech R&D, Buttel's caution seems reasonable. Wright (1986) shows how, for example, the development of genetic engineering offered new opportunities for growth in the petrochemical and pharmaceutical industries which while experiencing a rapid growth relative to other sectors needed to develop new technologies to offset the costs of the oil price rises in the 1970s. At the same time, DNA manipulation became politically more attractive to companies after the election of Reagan as US President in 1979: his administration lowered the regulatory constraints on and barriers to genetic research while raising incentives (such as tax-cuts and changes in patent laws) to encourage investment in the new technology.

While these circumstances led to a rapid rise in corporate activity in the area, Wright believes that the principal players, the multi-national corporations (MNCs), were acting defensively, participating in order to ensure that their products would not be undermined by DNA innovation. There has been a large number of new biotechnology companies formed, yet many operate in specialist, niche markets or supply the bigger established corporations with products or techniques. Even the largest biotech company that has developed beyond research to being a major producer of genetically engineered products, the US company Genentech, has (in 1990) become effectively incorporated into the larger biochemical/pharmaceutical MNCs through merger and restructuring with Hoffman-La Roche. A recent US Office of Technology Assessment report (OTA, 1988), confirms Wright's earlier prognosis that the field is primarily of interest to industrial sectors experiencing problems in their long-term international competitiveness, acting as a replacement for declining industries rather than a revolutionary harbinger of new industrial structures. It also records how important state funding is for the development of the R&D infrastructure: while US industry is currently spending about $2000 million annually on R&D, federal and state support nears $3000 milllion each year.

The revolutionary potential of biotechnology industry is then, uncertain. If there is one way in which this may appear it is through its capacity to alter the relationship between raw materials and the manufacturing process. Sargeant (1984) has claimed that biotechnology initiates a trend towards the 'dematerialisation' of production, that is, the emergence of smart (biological) products that not only need less raw material input for any given level of output but are also more powerful in what they do and so need only be bought in smaller volumes. In theory, single production plants could produce sufficient output for global demand of specific products. This is already happening in some areas: for example, the largest independent British biotechnology company, Celltech, makes over 40 per cent of the world's production of monoclonal antibodies for diagnostic and therapeutic use. Nevertheless, there is still a long way to go before the dematerialisation of production actually occurs. Corporate R&D may produce more powerful pesticides, sweeteners, hormones or whatever than in the past but they will want to ensure that their sales income is not thereby reduced: it might turn out to be more difficult to sell fewer at a higher price. Moreover, state regulatory authorities are taking more interest in the safety of, as well as necessity for, these new products (such as the growth hormone bovine somatotrophin which has been developed by Monsanto to increase the production of milk in cows). Both economic and political factors may well intervene, therefore, to limit the degree to which smart products are actually produced and marketed.

Finally, biotechnology can be said to have involved a 'revolution' in the relationship between science, technology and society. It has posed, and answered the question of whether life processes and organisms be *owned* as the patented intellectual property of individuals or corporations? Since 1980, the answer to this question has been 'Yes'. In June 1980 the US Supreme Court determined that existing patent law allowed the patenting of life forms that were 'not nature's handiwork'. This was in response to a patent filed by a scientist, Ananda Chakrabarty, working in the General Electric Company labs, who had created a special *Pseudomonas* bacterium. More recently, in 1985, the US Board of Patent Appeals decided that plants could also be patentable though this is not permitted in European patent law where new plant varieties are protected by plant breeders' rights Acts. However, plant material that contains genetically engineered characteristics that may be of use, such as disease-resistance traits, will be patentable. What is and is not patentable has, then, changed as the courts have tried to keep up with the novelty claims of the biotechnologist. Today, the constraint of prior existence in 'nature's handiwork' no longer prevails: 'Patent law has become much more liberal on this point by taking into account the merit of first discovery of a natural substance, its isolation and provision of a product in a form useful for therapeutic or other purposes' (Crespi, 1989).

There are, nevertheless, ethical questions about the ownership of novel life forms. There has been considerable debate about the research currently being conducted on the human genome, the complete sequence of genetic information carried in a person's DNA: could any of this information or work that it leads to be patentable, and if so, for whose benefit? In fact, the European Parliament has declared the human genome research strictly non-patentable.

While the human gene remains (as yet) uncommercialised, the same cannot be said for plant germplasm or seed. Large agribusiness companies have developed an increasing control over gene banks vital for the development of new plant varieties, either by traditional or biotechnologically informed breeding methods (often a mixture of the two). The seeds for many important crop species (cereal and root crops) have their origins in the Third World, but less developed countries are concerned about both their access to seed controlled by private companies and the cost of purchasing patented seed carrying genetically-engineered properties (see Kloppenburg, 1988). It is for these reasons that many advocate the maintenance of publicly-funded plant breeding and a biotechnology which is explicitly geared towards social welfare rather than commercial gain (e.g. Moser, 1988).

The concern over patenting life forms has been generalised into a wider debate over the direction and control of biotechnological research, most especially in the US through the political activism of Jerry Rifkin, a radical critic of corporate influence in this area. Perhaps more than any other area of scientific controversy, biotechnology has created a more politicised and increasingly informed public prepared to scrutinise and question the priorities of commercial R&D. This concern has arisen in part as a response to the issues raised by patenting, but also more generally as a result of anxieties over the safety and toxicity of genetically engineered organisms released to the environment (Tait, 1989). This issue of control over science and technology, not only in biotechnology but also elsewhere, will be explored in Chapter 6.

I have suggested that biotechnology has been exploited commercially because it has provided new technological developments which potentially will help sustain the commercial profitability of the large biochemical and biomedical industries. Its 'revolutionary' capacity to do more than this is less certain, despite the rhetoric that often accompanies biotech literature. Nevertheless, it has brought about important changes in the institutional processes through which science and technology are exploited within both public and private sectors.

Conclusion

I have tried in this chapter to provide a sketch of some of the key features of the historical and contemporary exploitation of science and technology by private capital. The chapter began with a discussion of the circumstances that led to the establishment of in-house R&D capabilities within companies and new increasingly sophisticated and flexible systems for the development and use of scientific expertise. Increased linkage with academia and rationalisation of the management of R&D are two important features characterising this process.

It is important to remember that the research-intensive companies are primarily the large multinationals, since it raises questions about the relationship between their *multi*national interests and strategies and the *national* interests of those states within which they operate. The more international the planning for corporate R&D (and production) the more difficult it becomes for individual nation states (particularly those in which the corporation originates) to manage and control corporate activities on behalf of what can be regarded as the *national* economy. Perhaps it is in response to the problem of balancing these potentially conflicting interests that policies for science and technology – both their exploitation and control – have themselves become more internationalised, especially in Europe.

More generally, I have suggested that though the sociological analysis of corporate R&D has been limited both theoretically and empirically, it promises much in terms of improving our understanding of

(i)   the social relations of scientific work in company laboratories;
(ii) the relationship between this and its distinction from other forms of labour used in the production process;
(iii) the organisational culture of corporate science and the way scientists as 'calculative actors', as we saw with Fujimura's account, respond to the specific demands of their employers

and negotiate research agendas that satisfy their scientific interests.

We saw in the previous chapter the way in which the state has played a key role in the exploitation of science and technology bases within specific countries. This has typically – though not exclusively – been geared towards the support of national capital. Stoneman (1988) has shown how this has been vital to the maintenance of the 'dynamic efficiency' of economic growth: he challenges the conventional view that 'the market' alone will be an efficient allocator and user of R&D resources, arguing instead for further state support for, and management of, the science and technology base. This still leaves open the question of whether the 'exploitation' of science and technology should be solely geared towards the market and towards profit. Many have argued that in the case of biotechnology social welfare priorities have taken a backseat to the commercial interests of those in the field. As Yoxen (1983) has argued: 'What we have to decide is not what institutional arrangements will keep the research front moving forward, but what institutional arrangements will allow research for social needs defined in other ways than by the operation of the market. How do we open up space for alternative programmes of research that set aside and confront the priorities of profit-maximising industrial corporations . . .?' (p.97). The possibility of controlling and shaping science in different ways is the subject of Chapter 6.

6 Controlling Science and Technology: Popular and Radical Alternatives

We should be on our guard not to overestimate science and scientific methods when it is a question of human problems; and we should not assume that experts are the only ones who have a right to express themselves on questions affecting the organisation of society.

Albert Einstein (1949)

For 'undue public concern' one should always read 'perfectly reasonable terror'.

Lucy Ellman (1990)

Controlling science and technology is an exceptionally difficult task. Who or what should one control: scientists, their institutions, wider technological systems? And for what purpose: what direction are we to take, whose interests are we to serve? And since, as we have seen, science and technology are malleable, socially negotiated, institutionally located and without clear boundaries, then what is '*it*' that we are trying to control?

Answers to such questions are intimately bound up with the question of *who actually controls science*? Those who do so are very likely to determine just what needs to be controlled. We are then concerned here with the politics of knowledge in a way which is much broader than that typically associated with the arena of formal science policymaking, though it will include this as well. We have to broaden our perspective to consider the power of corporate, professional and bureaucratic elites whose own priorities can set R&D agendas without reference to the forum of democratic policymaking.

At the same time, science and technology can and have been shaped and switched down different tracks through the actions of (non-elite) pressure groups agitating over specific issues as well as through the more sustained and fundamental challenges of alternative science movements and critiques.

Groups that seek to influence the direction of science have more or less effect depending on the political culture within which science and technology debate is located. As we saw in Chapter 4, for example, the more pluralistic and open the political culture, the more possible it is for interest groups to participate in a debate about science even though this can mean it is gladiatorial in style and resolved through litigation in the courts. Whatever the particular political culture there is also the question of how scientists and particularly their elite members – senior members of the principal professional organisations and elite research institutions – respond to the call for a wider public involvement in the direction and priorities of science. Frequently, one hears the argument that the public is insufficiently informed about science to be able to make 'sensible' judgements about areas of research. For its part, the public might in fact accept the wisdom of science and seek only to be informed in order to comment on areas of concern.

The sort of public concern that is expressed over science and technology clearly depends on the image and understanding the public has of it. In Britain, this question has been explored through a research programme formally dedicated to analysing 'the public understanding of science'. There is a number of particular projects being pursued as part of this programme. Most importantly, the work assumes that there is a *variety* of publics, not one undifferentiated mass, each having its own specific level and type of understanding. One of the studies has involved a national survey of public attitudes towards science. Such studies are useful in providing an overview of public opinion, sources of information and levels of knowledge about science itself. But they have their limitations, especially in charting in a more qualitative fashion the way people experience and regard the impact of science and technology on their daily lives. This is of vital importance since such experience is more likely to shape the way people translate their attitudes and concerns about science into social action. It would then be possible to explain why some people are more predisposed to seeking out information about science and acting on it while others exposed to similar sources of information remain relatively unconcerned.

We need, too, to distinguish between people's understanding of the *content* of science from their knowledge of it as a *social institution*. Yet institutions of establishment science, such as the British Association for the Advancement of Science or the Royal Society, normally focus on the first rather than the second of these two forms of understanding. For example, in its report, *The Public Understanding of Science*, the Royal Society (1985) suggests that: 'In a democracy public opinion is a major influence in the decision-making process . . . To decide between competing claims of vocal interest groups concerned about controversial issues such as "acid rain", nuclear power, *in vitro* fertilisation or animal experimentation, the individual needs to know some of the factual background and to be able to assess the quality of the evidence being presented' (p.10). As Collins (1987) has commented, however, while it is no doubt important that people understand more about science in this technical sense, it would be wrong to presume that they will therefore be able to enjoy a more 'objective' and 'authoritative' position with regard to science. As we saw in Chapters 2 and 3, disputes between scientists themselves about what is or is not 'good' evidence characterise all areas of scientific debate, most especially areas of controversy, such as all those mentioned in the extract from the Royal Society report above. Collins observes that 'Even among the experts themselves, who have been trained to many levels above what can be expected of the public's understanding, radically different opinions are to be found.' He concludes therefore '[i]t is dangerously misleading to pretend that the citizen can judge between the competing views of technical experts when even the experts cannot agree' (p.691). I shall return to this point towards the end of this chapter since it has important implications about the relationship between expertise, decision-making and accountability to the public.

The image of science portrayed by science is one of certainty and authority. Media representations usually work to confirm this image. Like other areas covered by the media, the media conveys not merely information, it also tells us what 'it thinks' is important. Because of this, it is often argued that the media does not merely record events in any neutral way, but instead is very selective in what it presents and how it is presented, tending to favour some interests over others. A key reason why the media can be so effective in shaping our ideas is because it deliberately builds on pre-existing cultural stereotypes or beliefs of what is right or wrong, for example in connection with industrial relations or race affairs: the journalist or editor will draw on these beliefs to frame or even 'explain' a story, even if this is used in a totally unjustified way.

The image of science presented in the media has, then, until fairly recently, been one which has stressed its authority: this conforms to the widespread cultural belief we have about science and which scientists themselves will try to promote. Advertising agencies can happily draw on this stereotype in order to sell new products. As Jones *et al.* (1978) comment (in one of the few detailed British studies of the relationship between the media and science): 'The tendency [of advertisers] is to juxtapose a consumer product, or consumer-oriented advice of all types of authenticity, and a stereo-typed image of science. (An obvious example of such an image is the ubiquitous white coat.) This use of science testifies to the very powerful role allotted to science in legitimating information . . . The actual explanatory value of the feature articles and advertisements is often low: it is basically the symbol of scientific authority and impartiality which is being invoked' (p.6). A similar sense of the authority of science is typically conveyed in news or documentary coverage about science. Until recently, science has been presented in a rather superficial, uncritical way: either the naive and reverential 'gee-whiz, isn't that amazing!' approach (of British programmes such as *Tomorrow's World*) or the simplistic and somewhat patronising commentary on what this 'discovery' will mean for the 'ordinary man (*sic*) in the street'.

Even while this sycophantic and shallow approach to science still informs much of the coverage, over the past decade science issues have been approached with a more critical eye because of the very politicisation of science itself. The old stereotypes of certainty and authority look less suitable frameworks within which stories about the environment, nuclear power, genetic engineering or whatever can be located. In all these (and many other) areas, 'the public' has become increasingly involved in examining and challenging scientific expertise, as we have seen. The more people become involved in scientific debate – via pressure groups, political parties, public inquiries and so on – the more difficult it is for the media to package science in simplistic stereotypical images precisely because such images only ever work where people have a partial, distant understanding of the relevant area.

However, even when controversial issues in science are covered in greater depth, there is still a tendency to portray the problem as one which can be dealt with by either better science or, more usually, a better application of science (and technology). There is rarely, if ever, coverage which seeks to problematise the nature of scientific inquiry itself, which might show its constructed and negotiated and uncertain character. Indeed just the opposite is the norm, with major 'serious' science programmes depicting scientific exploration as the asocial, disinterested pursuit of 'the truth'. Moreover, the regulation and control of science is usually presented as a matter best dealt with by the internal, self-policing mechanisms within science, such as peer review, ethics committees, licensing authorities, and so on. Of course, the 'audience', just like 'the public' is neither uniform nor uncritical. Moreover, the message the media intends to put out is not necessarily that which people pick up. There is a wide range of interest groups that are highly critical of the conventional view of science as paraded in the press and TV, precisely because it tends to serve the interests of establishment science and technology. For such groups, there is a need for an *alternative* approach to science, to reflect the priorities of those who are not part of the establishment. The next section presents some of the more important of these voices for an alternative science and technology.

Alternative science and technology

We need to distinguish between two forms of critique of science: the short term, issue-based activity of *pressure groups* that come together to challenge direct or indirect effects of technological change, and the long-term alternative science *movements* that seek a fundamental revision to the character and direction of science and technology. The distinction between these two is not always clear: for example, pressure groups can become transformed into wider and more long-lasting organisations, such as the Friends of the Earth, which began in 1970. In turn, these can become part of a wider socio-political movement, such as we have seen with the growth of environmentalism over the past twenty years. Nevertheless, I shall examine some of the principal features of pressure groups first.

The role of pressure groups

Pressure groups protest about specific issues such as the siting of a proposed airport, the dumping of nuclear waste, the hazards associated with new drugs or the development of new research procedures. Typically, they are those most directly affected by such matters, and often combine people from a variety of social backgrounds though the more articulate and politically powerful middle classes tend to orchestrate the protest action. As Nelkin (1984) has noted, protest groups are much more easily formed around a specific issue since this tends to generate its own 'natural' constituency of activists. The development in Britain of the nuclear power plant at Sizewell is a good example of this. The less concrete, less focused the issue is, the more difficult it is to create a strong sense of shared interests that can be translated into effective action. So those who might be prepared to challenge the development of the nuclear plant at Sizewell may be less prepared to join wider protest movements against nuclear power itself.

It is important to remember that pressure groups may not always be politically oriented towards the left. Many times one sees groups challenging technological change because it is seen to threaten traditional or conservative values: the pro-life anti-abortion group is a case in point. Another is the 'Creationist Science' lobby in the United States which has advocated the removal of Darwinian evolutionary theory from biology courses in American schools on the grounds that it is scientifically not proven and contrary to the Christian belief that 'man' was created by God. Interestingly, creationists are not 'anti-science' since they seek to gain support for their own 'alternative' history of the world through what they regard as a conscientious adherence to the scientific method: hence the name 'creationist *science*'. Moreover, they appeal to the American political tradition of pluralism by arguing that schoolchildren should be exposed to a variety of explanations, not one (the Darwinian).

Sometimes, both right and left may ally themselves together in concerted action against what each regard as potentially threatening developments in science: here, for example, the anti-vivisection pressure group contains those on the left who oppose the use of animals in medical research on the grounds that the experimental work is designed to produce yet more 'me-too' drugs for the highly profitable pharmaceutical industry, while those on the 'right' may regard animal experiments as contrary to a 'man's traditional relationship with animal life and nature'. Anti-vivisection is, in fact, a good illustration of the way in which differing interests have been woven together historically, interests which have not always been primarily motivated by the welfare of animals *per se*: as often as not they have reflected the social interests of the groups involved in the critique of vivisection.

French (1975) shows, for example, that the Victorian establishment was hostile towards vivisection because it epitomised the emergence of the new scientific medical research and practice whose members posed a professional threat to the then dominant status groups within British society. Rupke (1987) makes a similar point saying that the aristocracy, clergy and judiciary 'saw their cultural influence waning . . . [as] the new generation of professional scientists sought neither ecclesiastical preferment nor aristrocratic patronage'; as such 'they represented an encroachment on traditional estates of cultural authority' (p.8). 'Establishment' rather than animal welfare seems to have been important.

Commenting on this analysis of the social basis to the antivivisection lobby, Yoxen (1988) emphasises the importance of seeing the growth of pressure groups as a dynamic process involving negotiation and coalition between the different interests that make them up. While there has been a considerable amount of research on the formation of groups in this way, there has been relatively little sociological analysis of the sort of responses made by those working within science to the challenge posed by activists. In the case of vivisection, then, '[i]t would be interesting to know how research scientists, industrial companies dependent on the use of animals in research, and organisations that trade in laboratory animals acted in recent years' in response to the growing animal welfare lobby, including its more radical wing, the Animal Liberation Front (ALF). 'What alliances did [the scientists and others] form? What supporters did they seek to enrol? What were their perceptions of the changing political environment? Which groups tend to be more receptive to the idea of alternative modes of testing and/or to stricter laboratory standards?' (p.45).

Animal activists are not 'legitimate whistle-blowers'
Pressure groups are allowed to exert pressure but only within certain limits, as the following extract about US activists shows.

A growing number of break-ins, fire-bombings and threats by animal-rights extremists is fuelling movement in Congress towards a law that would make crimes against animal-research laboratories a federal offence.

Claiming that state and local police forces are not competent to prevent animal rights attacks on research facilities, several members of Congress have recently introduced bills that would allow the Federal Bureau of Investigation (FBI) and other federal police bodies to investigate crimes by animal activists.

The legislators argue that assaults on animal research laboratories are often planned and executed by national animal-rights organisations, and that participants often cross state lines. They note that at least one extremist organisation, the Animal Liberation Front, began in the United Kingdom and continues to have ties there. In the past decade, illegal acts by animal-rights organisations have increased more than three-fold . . .

Testifying at a hearing on one of the bills last week, Richard Van Sluyters, from the University of California, Berkeley, who has been targeted by animal-rights groups, characterised the activists as 'professional agitators . . . Their interstate conspiracy is beyond the local law-enforcement ability . . . When you tell a local policeman that someone has taken your rats and written on your walls, they don't understand the level of tragedy.'

Source: *Nature* 15 Feb. 1990

It is important to note, however, that the political arena in which pressure groups operate is only one forum in which politics is, as it were, 'made'. That is, the open, negotiated and gladiatorial struggle characterising such interest groups, is only one medium through which power is expressed. Alongside, there is the less open, less available arena of power where one finds elite decision-makers – government officials, politicians, the economically powerful – setting political agendas *in advance* of debate in order to limit the terms of discussion to those that are considered to be 'manageable' within the context of the status quo (see Lukes, 1979).

Radical activists, like the ALF, justify their militancy precisely because of their cynicism towards the normal procedures of government and orthodox politicals being used to 'fix' debates in this way. There is evidence that this does occur, especially through the way in which the mechanisms for debate are sometimes set up. For example, when, in 1982, the US drug company Upjohn was refused a UK licence for its contraceptive drug, Pepo Provera, the company appealed against the decision. The appeal Panel then met to determine who would be allowed to give evidence. Groups (especially feminist groups) opposing the drug for its alleged carcinogenic effects found that they were only allowed to submit evidence in writing and only through Upjohn itself: as Lewis (1986) comments, 'This arrangement served to prevent opponents actively contesting claims made at the hearing, and also provided the applicant company with prior knowledge of their criticisms' (p. 46).

In part as a response to perceived inadequacies of establishment politics and the 'due political process', it has become increasingly common to find pressure groups forming their own organisational structures through which debate about the impact of science and technology can be conducted. This poses a threat to orthodox science institutions whose members, quite simplistically, tend to ascribe this interest in alternative groups as a symptom of the wider public's failure to understand the new technological developments: hostility is based on ignorance. In fact, studies have shown that people are much less resistant to technology *per se* but are more disturbed by the lack of opportunity to participate in decision-making about science: as Ince (1986) has argued, 'if the decision-making system is opened up, and effort made to reduce the enormous distance between our scientific institutions and the people they affect, this will be a better inducement to make an effort at understanding what is at stake than any amount of well-meaning, if self-interested, exhortation' (p.191).

However, while we might agree with Ince's proposition, it is important to recognise that the terms on which wider participation might be offered could be far from genuinely democratic. This is the point stressed by Dickson (1984) who has distinguished between 'public participation' in decision-making and a more substantial 'democratic control' over science and technology. The first is based on what he calls a 'technocratic approach' to science policy that involves consultation with interest groups but decisions determined by a consensus among the (technical) experts: 'rule' by a technocracy. The second, presupposes that 'solutions to the problems caused by the applications of science require a redistribution of political power as much as the insight of technical expertise' (p.219). Dickson believes that the evidence suggests that the first of these two approaches is currently dominant, especially in the US science policy arena, and, echoing my remarks above about agenda-setting by the more powerful, he notes that they can 'lay down the boundary conditions for participation, determining which types of argument will be considered by decision-makers, and thus defining the limits of legitimacy in both technical and political terms' (p.220). It is for these reasons that alternative science *movements* have developed in the past and present, seeking a fundamental redistribution of power in the way science is determined, which presupposes, of course, a fundamental change in the wider social system. I want now to look at some of them, focusing attention on 'appropriate' science and technology, radical science, and feminist science.

Alternative science and technology movements

Appropriate science and technology

The idea of an 'appropriate', or sometimes, 'intermediate', technology can be traced back to the efforts of Fritz Schumacher. He gained experience as an engineer and adviser in India trying to establish local mining and manufacturing plant during the last few years of British colonial rule, and decided that it was against India's interest to import expensive technology from Britain in order to modernise its economy. Expensive, capital-intensive, difficult to maintain and wholly inappropriate to what was (and still is) a preponderantly *rural* economy, Western technology needed to be replaced by machinery and a productive enterprise that could be locally sustained, a more 'appropriate' technology not only for India but for the whole of the developing world.

Subsequently, Schumacher published his most famous text, *Small is Beautiful* (1973), in which he explored the possibility of developing a technological system for the masses: cheap, easy to maintain and labour-intensive. Local, small-scale industrial enterprises, he said, should be developed using local resources and skills and geared towards local markets. As a result of his work, the Intermediate Technology Development Group (ITDG), based in Rugby, England, became a centre for the development of alternative technology (AT) projects, with a special though not exclusive emphasis on Third World countries.

During the late 1960s and early 1970s, Schumacher's ideas were well-received by a generation of younger people, especially among the middle classes, unconvinced by the industrial and political promise of an industrial capitalism, that offered only increasing militarism during and after Vietnam, growing unemployment and a scarred and polluted landscape with urban dereliction to match. Science and technology had been ensnared by the military–industrial complex and could no longer be trusted to answer the growing social and cultural uncertainties that people experienced. However, Schumacher was never a prophet for the ensuing counter-culture that developed among the young: prophets came from the world of music, fashion and philosophy.

All this is now some distant memory as counter-culture has become marginalised and replaced by a more atomised and privatised ethic where the language of social justice and the critique of capitalist technology is more difficult to learn, to speak and to convey to others. In many ways, of course, what was the counter-culture of the 1960s and 1970s has since become more respectable, especially in the form of environmental awareness. The 'environment', a rather ill-defined concept evoking an image of 'Nature' beseiged, is now a central political and economic issue. It is, however, one which in practice rarely challenges the existing political and economic status quo. Unlike the radical language of the 1960s, environmentalism has been captured by those in power.

Lowe and Goyder (1983) believe, for example, that the pressure to protect 'the countryside' typically comes from those in a position of power defending their own 'backyard' or green acre against expansion from elsewhere. Newby (1987) also argues that the ideology of 'our national heritage' can be used very effectively by those in power to defend the 'environment' which *they* enjoy compared with the 'degraded environment of the deprived'. Within the formal political arena, established political parties have had to ensure they portray a 'green' image to the electorate, as well as to incorporate or ally with the increasingly strong 'Green' parties, especially in Western Europe. Inasmuch as this has happened, then the extent to which environmentalism needs to be anti-industrial, anti-growth, or indeed anti-capitalist, appears to decline. This is reflected in the social basis of those expressing a concern for the environment: Inglehart (1987) suggests that the traditional class politics associated with the period of industrial expansion and growth, a politics which generated both pro- and anti-capitalist forms of organisation, discourse and practice, is being replaced by a post-industrial politics where forms of collective action are less closely tied to the conventional concerns of distinct social classes. Thus, Green Party politics have been said to be radical in a 'post-Marxist' sense (Boggs, 1986), and as Papadakis (1988, p.448) has noted: 'Although located on the left of the political spectrum, Green parties generally diverge from the traditional Left through their focus on non-economic goals. Their concern with a far wider range of issues and their broader vision of the future has reinforced the view that the traditional Labour, Socialist and Communist parties' are no longer viable.

While an 'alternative' anti-capitalist discourse may now be more problematic to sustain, (especially after the media rhetoric associated with the decline of state socialism in Eastern Europe), the practice of promoting appropriate technology around the globe continues today through the efforts of both First World organisations like ITDG and Third World movements, such as the rural co-operative *Proshika* movement in Bangladesh. Moreover, within industrial countries there has developed a number of self-help, AT groups. In Britain, for example, the Centre for Alternative Industrial and Technological Systems (CAITS) grew out of an AT initiative by workers at the Lucas Aerospace Corporation to develop socially useful products and provide full employment. The work of CAITS received further support from the Greater London Enterprise Board, which was part of the late lamented Greater London Council, but now operating as an independent agency supporting local and regional technology networks inspired by the AT philosophy. In the Netherlands, 'science shops' were established during the mid-1970s by the Dutch radical science movement. They provide a community-oriented scientific advisory service, linking and liaising between citizen groups and more distant scientific experts, especially those in the universities.

These initiatives have had some limited success in redirecting science and technology down what many regard as socially useful paths. However, though the ITDG argue that 'small' is not only 'beautiful' but actually 'possible' (McRobie, 1980), its impact is likely to remain small in the more conventional sense of the term. This is because of the failure of AT to address fully the question of the political economy of production systems: without confronting the basis of power and class within capitalist economies in either the Frist or Third Worlds, AT will remain a relatively marginal (even though valuable) adjunct to the predominant (capitalist) system of production. Just as environmentalism has in some countries been successfully incorporated within the existing status quo, so the language and practice of AT has posed little threat to existing social structures. Where it has done, those in power will seek to co-opt it.

AT practitioners have, then, typically ignored the social relationships of technology, concentrating instead on the *products* (like a 'bio-gas reactor' or underwater turbine) they can devise for appropriate clients. Because of this, many AT projects fail to achieve their longer-term objectives. In the Third World, for example, the introduction of tube wells as cheap and easy-to-maintain devices has not made access to water much easier for the poorer farmers who were supposed to be the main beneficiaries of the new technology. Instead, precisely because the wells represented a threat to the richer farmers, the latter have ensured that it is they that control the technology and they who sell to the poorer peasantry the greater supply of water the wells have brought. As with other forms of technology associated with the so-called 'Green Revolution' in the Third World, the more prosperous agricultural households are the ones most likely to benefit from and control the introduction of new farming technologies, whether they are 'appropriate' or not: the technology becomes incorporated into the local social hierarchy in such a way that, in effect, it becomes essentially a landlord-biased technology. Thus, those who criticise indigenous farming populations for failing to take up new AT opportunities could be accused of ignoring the way in which the local social structures prevent this happening. As Cromwell (1989) has argued: 'The political dimension, perhaps the most important factor determining technology-related policies, is frequently neglected by advocates of AT in favour of micro-level technical, economic or social factors. Failure to adopt a given technology does not imply passivity or apathy but arises from the structural arrangement of the economy' (p.203). Moreover, Cromwell believes that, typically, AT projects are supported by Third World governments and international aid agencies as 'welfare' measures rather than as ways of contributing to genuine economic development: 'Currently AT is widely used as a welfare measure and until it becomes part of mainstream economic development, rather than simply a tool for achieving the social and political objectives of policy, AT will remain marginalised in many countries' (p.209).

From the economist's perspective, even if AT were to become a more important dimension of economic planning, it is not clear that it would generate sufficient economic growth to provide enough jobs and consumer goods for a growing population. While AT favours labour-intensive policies for the jobs they create, an *expansion* in employment is much more difficult to achieve: as Clark (1985) has said, 'even if labour-intensive techniques do exist and could be adopted, policies promoting them would lead to technological stagnation since the potential for improving on them is strictly limited' (p.187).

Sociologically, the lesson to be learned here is that defining 'appropriateness' simply in terms of some technical criteria – such as ease of maintenance – is insufficient: we need to incorporate into our definition an understanding of the way the technology will be '*appropriated*' by those in powerful positions. At the same time, there is a danger of forgetting to consider the way in which those in more subordinate positions modify or adapt new technologies to fit in with the pre-existing technological infrastructure. Some social groups are more able to do this because of the class position they occupy. Settled owner-occupied farmers are, for example, in a much better position to adapt and take advantage of AT or other technologies that appear than are, say, nomadic or pastoral farmers moving from place to place in search of new grassland. At the macro-level, then, certain types of technology are linked to and shaped by certain types of social relations. It is this idea which has inspired the 'radical science' movement, whose advocacy of a more appropriate science and technology goes beyond the (worthy but) limited ambitions of the AT movement. Their call for a 'science for the people' is premised on a fundamental change in the basis of society.

The radical science movement

As a social movement, radical science has had its ebbs and flows. This is to be expected when any social collective lacks a strong institutional and organisational basis, and when its fortunes depend on the strength of charismatic leadership and focal issues around which members can unite. Radical scientists have tried to establish different forms of organisation to strengthen their position, such as the Association of Scientific Workers formed in Britain and the United States, very active during the 1940s, and the British Society for Social Responsibility in Science formed in 1968. Such organisations helped to build up networks of concerned scientists within both the natural and social science disciplines.

There have been two periods during which radical science has been most active: the first covers the two decades of the 1930s and 1940s, the second the late 1960s through to the end of the 1970s. Both periods of activity were initiated by scientists' growing hostility towards the use of science and technology for military purposes, the first, in connection with the atomic bomb, the second the war in Vietnam. The horrors of Hiroshima or napalm raised fundamental ethical and political questions about the use of science by the state.

During the 1930s and 1940s, a growing number of European and US scientists came to regard science and technology as misdirected and abused by those in power, to control and dominate nature and society – but especially the working classes – in general. Rather than solving the problems of the world – such as the prolonged recession and warfare – science seemed simply to make them worse, used by non-accountable 'experts' to justify inequalities, destruction and authoritarian state control. One of the most important books to appear at this time was John Bernal's *Social Function of Science* (1939). Bernal's thesis was that the organisation of science within a capitalist, ideologically bourgeois democracy was detrimental to both scientific progress and public welfare. According to Bernal, the growth of the scientific tree of knowledge is unnaturally stunted by the economics of capitalism, so that 'science is applied when and only when it pays' (p.128). Science could only function properly when founded upon a socialist economy, where service rather than profitability came first. Bernal's ideas were quickly taken up by influential sections of the international scientific community, many of whom looked towards the Soviet Union for the seeds of a liberated and liberatory science (see Kuznick, 1987; Werskey, 1978).

The enthusiasm for a 'socialist science' in the East was, however, soon to be dampened by the postwar Stalinist regime which gave little or no room for the sort of democratic science many in the West had hoped for. The 'Lysenko affair' was a further blow to the socialist aspirations of Western scientists: Lysenko was made director of the Institute of Genetics after winning the support of the Soviet authorities over his claim that acquired characteristics of plants are inherited, a direct attack on the conventional Mendelian genetic theory which most geneticists both in the West and East accepted. Lysenko's ideas became part of the general assault on 'bourgeois' ideology and science – which led to some geneticists ending up in Soviet labour camps – and supported Stalin's break-neck pace of industrialisation and agricultural development. While Lysenko's ideas were later discredited within and outside the Soviet Union, they have to be seen in the context of Stalin's desire to force the pace of cultural change in the country towards his image of the genuinely proletarian society, in a way not unlike the Cultural Revolution in China during the 1960s. Nevertheless, radical scientists were dismayed and disillusioned by Lysenko for it seemed that Marxist theory had corrupted scientific practice.

After a prolonged period during the 1950s and 1960s of withdrawal from the political arena, radical science re-emerged in the late 1960s inspired by the critique of capitalism and imperialism which had been catalysed by the Vietnam war, by Third World independence struggles and by the growing attraction of Maoist models of popular socialism, which seemed to avoid the terrors of Stalin's version (even though, in reality, there were many in the interior and the margins of China – such as those in Tibet – who had felt the full force of the Chinese purges).

Elzinga (1988) has argued that the new radicalism of the 1970s was split into three, the moderate, the more radical, and the 'ultra-leftist' tendencies, whose principal figures are Jerry Ravetz (1982), Hilary and Steven Rose (1976), Bob Young (1977) and David Dickson (1984) respectively. Ravetz challenges the view – one that can be found in Bernal's own work – that science is 'underneath it all' really a democratic and altruistic pursuit. Instead, scientists can be as corrupt, as opportunist, as elitist and sexist as anyone else. The point is to try to limit these tendencies where possible and build on those opportunities that enable science to become more accountable to society at large. The Roses adopt a more radical position than this by arguing that science is not simply bad science when it is 'abused' by its practitioners: the very nature of scientific inquiry in capitalist society embodies exploitation and class-based ideologies within its very ideas and theories, whether they are about IQ and intelligence, racial differences or gender and biology. Young and Dickson take this argument a further step by claiming that we can and must distinguish between 'capitalist' and 'socialist science' providing us with different theories and facts about nature and society. The science in each is no more or less true than the other: they simply serve different purposes and social classes, though, of course, inasmuch as proletarian science aims to serve society as a whole, in practice, it provides the model to be achieved.

As Elzinga notes, since the 1970s 'these [three clear] positions broke apart, and there has been a fragmentation into single-issue approaches', and he asks 'the interesting and important question that comes out of all of this is: how can we maintain an approach that is critical, but at the same time resists the temptation towards fragmentation?' (p.113). His answer is that we should not cut ourselves off from the past but learn from it. Perhaps this would help radical science itself to recognise what has been – despite the rhetoric – relatively ignored in the debates about science and capitalism: this is the need to give much greater attention to the relationship between gender and science. As Hilary Rose (1987) herself comments more recently: 'Looking back over the writing of the sixties and early seventies, it is difficult not to feel that, as the critical work became more theoretical, more fully elaborated, so women and women's interests receded. Thus the writing gives no systematic explanation of the gender division of labour within science, nor, despite its denunciation of scientific sexism, does it explain why science so often works to benefit men' (p.274). It has been feminist sociologists rather than radical sociologists of science who have done more to establish the credentials and critique of the third movement I want to consider here, 'feminist science'.

Feminist science

It is rare to meet many women at a science conference or convention. Usually the participants and speakers are male: there are relatively few discos at science conferences (apologies for this heterosexist remark). Yet there are many women who work – in education, industry or the public services (such as hospital) – who have considerable scientific training and who are closely associated with 'laboratory life'. This is a life, however, that for most involves fairly routine laboratory skills servicing more senior science staff. It is exceptional to find women lending research teams as principal investigators, and where this occurs it is typically in the biological field to which a disproportionate number of female recruits to science are drawn. Even when women excel in their particular field of work, they have often failed to receive recognition for their work.

Feminist historians, philosophers and sociologists of science have explored the specific question of 'Why so few female scientists?' as part of a wider investigation into the relations between gender and science Schiebinger (1987) has suggested that there are four main dimensions to this feminist analysis: an historical one that opens a new, neglected history about women's long-standing involvement in science; an institutional one that charts the place and number of women in science today; one that challenges the image of women portrayed by natural science itself; and finally, one that poses a challenge to the very form and content of 'the scientific method' as being 'masculinist' in both theory and practice. Related to this last point, though not dealt with by Schiebinger, is the way in which technology and its development and use have been dominated by men, who have determined the pattern of technical change and the needs it serves. This case has been argued most strongly with regard to the development of reproductive technologies.

Women's place in the formal history of science is notable for its absence. Any science encyclopedia is likely to list relatively few women deemed to have made major contributions to science and engineering. Most role models of science – especially for the schoolchild – are male, such as Newton, Faraday, Einstein, Rutherford and so on, while the female image is catered for by just a few well-known figures such as Maria Curie and even, more tenuously, Florence Nightingale. Many of the recent feminist histories of science have sought to correct this neglect by producing biographies on a number of female scientists such as Mary Somerville, Hertha Ayrton and Rosalind Franklin. While this work has been very important it serves to throw into relief what for many women has been the typical experience of working in science, namely serving as technicians or assistants in a subordinate role.

Moreover, just as the issue of gender only became central to mainstream sociology during the late 1960s and onwards, so in the sociology of science earlier work by Merton and his colleagues in the United States or their European counterparts had ignored the question of women's place in science: as Schiebinger (p. 14) notes, while Merton 'pointed out that 62 per cent of the initial membership of the Royal Society was Puritan' he failed to register the fact that not a single one was female. While sociologists now attend closely to gender subordination, the actual situation for women in science has improved only slightly: in most science areas women are disproportionately scarce in the upper ranks and prestigious, secure positions of educational and research institutions. In science and engineering faculties, for example, women only account for 9.0 per cent and 2.0 per cent of staff respectively. Griffiths comments on the wider social constraints that restrict the recruitment of more women into engineering (see the boxed text below).

Women in engineering

Women's relegation to the low-status, low-paid and low-skilled jobs in the engineering industry means that their participation in practical shop-floor work is seen as very undesirable. The female apprentice has to be prepared to enter an almost totally male world. It is a world in which women are allowed only two roles – sexual object (the nude calendar, the pin-up photograph) and domestic support (girlfriend,

wife, mother). There is no room for a woman as co-worker, colleague or boss. The girl who makes this choice is likely to feel isolated. She will probably also be seen as a threat to her peers and supervisors who, not knowing how to respond to her, may resort to ridicule and belittlement. Because she is trying to do something different which, by definition is not women's work, it will be said that *clearly* she can't do the job as well as a man, can't be serious about doing it, can't be a 'proper woman' for wanting to do it, and so on. In the face of all these barriers, I find it surprising not that only 0.3 per cent of craft apprentices are women, but that there are so many.
Source: Dot Griffiths (1985)

While women can now become members of the Royal Society in London, its first female 'Fellows' were not admitted until 1945; in Paris, women were to wait until 1970 before they could enter the French equivalent, the Academy of Sciences. Such institutions have long been shrines to patriarchal values, without in any way being diffident about this: as the first secretary to the Royal Society, Henry Oldenberg, declared in 1664, its purpose was to 'raise a Masculine Philosophy . . . Whereby the Mind of Man may be ennobled with the knowledge of solid truths'. Even though there has been a steady improvement in women's position in science, in terms of seniority, levels of pay, career prospects and recognition for contributions, they are still disadvantaged relative to men in the same or equivalent position, as Cole (1979) has shown.

Apart from there being formal (sometimes legal) restrictions on women's access to positions of pre-eminence in science, there are much more powerful and effective cultural mechanisms at work which reduce female participation: as Rose (1987) has argued, laboratories are primarily 'Men's Laboratories' in which women work. Although ethnographers in the sociology of science have explored the cultural dimensions of 'laboratory life', its specifically gendered nature has been only properly examined by feminists. They show how the experience for many women in laboratory science is one of relative exclusion, isolation and a feeling of being given less opportunity to contribute to the experimental work of the lab and the direction it takes compared with their male counterparts. As Reskin (1978) has shown, in these circumstances it is more difficult for women to develop full collegial relationships with men.

Where women scientists have been successful, their biographies and autobiographies have been said to exhibit a 'sexlessness' (Rose, 1987) that hides the way in which gender has disadvantaged them. Many who are successful still depend on the patronage of men to help them with their careers: such women typically come from the upper middle classes whose relative affluence enables them to overcome the primary demand society places on them – that they be domestic labourers – by employing other women to do this work for them. Those not privileged in this way have to cope with the conflicting pressures to succeed in their jobs and to be the principal 'homemaker' and child-carer. Often this proves too difficult and many leave their jobs, to return, perhaps part-time, when their children have begun school.

Beyond the issue of the female participation rate in science lies the question of the masculinist bias in scientific knowledge and its technological forms. Sociologists of science within the 'interests' or ethnographic schools have explored in different ways the sense in which the very status of knowledge as 'scientific' depends on the negotiation and contest over ideas, which in turn reflect distinct interests and strategies. However, they have failed to chart the sense in which such ideas may carry interests or strategies which are gender-based.

Feminists have argued that the institution of science marginalises and exploits women: it does so not simply through the limited opportunities it holds out to them, but also through the methods, models and assumptions it adopts. Some feminists have sought to challenge what they see as the masculinist or 'androcentric' orienta-tion of science with an alternative 'female sensibility or cognitive temperament', as Delamont (1987) describes it, towards the explo-ration of nature (see, for example, Rose [1987] below). Not all feminists share this view, since for critics it simply substitutes one gender-biased approach with another equally as partial and so equally unacceptable (see Beier, 1986). It also presupposes that there is a specifically *feminist* methodology which provides a different access to nature and so the possibility of novel understand-ing about nature, which might in certain cases be more productive than alternative masculinist accounts. This is a position advocated by Evelyn Fox Keller (1985) who argues that masculinist 'valences' or impulses in scientific theorising and choice of theory tend to favour models of the world in which the principles of domination and aggression prevail: the DNA molecule is known as the 'master molecule' determining the very structure and process of life. Against this image, feminist geneticists have posed more holistic, less hierarchical models of the organism and the role of genetic structures within it.

> ### Towards a feminist biology . . .
> Unlike the alienated abstract knowledge of science, feminist methodology seeks to bring together subjective and objective ways of knowing the world. It begins with and constantly returns to the subjective shared experience of [female] oppression . . .[T]here is a sense in which theoretical writing looks and must look to the women's movement rather than to the male academy. Working from the experience of the specific oppression of women fuses the personal, the social and the biological. It is not surprising that, within the natural sciences, it has been in biology and medicine that feminists have sought to defend women's interests and advance feminist interpretations . . . [A] feminist biology does not attempt to be objective and external to the female biological entity; it attempts to make over biological knowledge in order to overcome women's alienation from our own bodies, our own selves.
> *Source*: Rose (1987) pp.279–81

The belief in an alternative feminist methodology raises a number of wider theoretical questions, however. As Richards and Schuster (1989) have argued, it appears to deny some of the now well-established principles of the sociology of science: most importantly, it appears at odds with the relativist proposition (see Chapter 2) that scientific knowledge is constructed via negotiation and competition in ever-changing contexts, and that therefore no set of knowledge-claims within science can or should be treated as in any way 'better' or more rational than any other. It implies that there is a 'better' feminist epistemology available that has been marginalised or denied a voice in science. At the same time, Richards and Schuster criticise Keller for reifying both 'feminist' and 'masculinist' influences on science: again, they would argue that the way gender shapes scientific debate is context-dependent, that the language and valency of gender is a discourse that is 'fluid and flexible'. This view has also been expressed by the feminist historian Jordanova (1989) who has explored the images of gender in science and medicine over the past three centuries. She points out that gender itself, and in particular images of womanhood and sexuality, have had unstable, fluid and continually constructed and deconstructed images over this period. Such historical evidence would mean that it is misplaced to identify *specifically* 'feminist' discourses within science counterposed to masculinist ones.

Nevertheless, feminists have shown how important the issue of gender is within scientific discourse, an issue which the traditional sociology of science had long neglected. Feminist contributions have been particularly strong in developing our understanding about the shaping of technology, especially in the medical and biological fields. There is now a large volume of literature examining the development and character of *reproductive technologies*, such as *in vitro* fertilisation, embryo research, and amniocentesis (see Stanworth, 1987; Corea *et al.*, 1985). Some of this writing is very hostile towards such technologies, seeing them as both male-controlled and threatening to women, as illustrated in the extract from Corea in the boxed text.

Reproductive technologies and female subordination

[W]omen must ask from whence [reproductive technologies] come. Why do these fabulous medical techniques require that women adapt to the most painful and debilitating circumstances? Why do such technologies reinforce the bio-medical 'fact' that a woman's reproductive system is pathological and requires an enormous amount of bio-intervention? Why do these techniques reduce the totality of a woman's being to that which is medically manipulatable? Under the cover of a new science of reprodution, how is the female body being fashioned into the biological laboratory of the future? And finally, will the ultimate feat of these technologies be to remove not only the control of reproduction, but reproduction itself, from women?

Others suggest that while reproductive technologies need to be challenged by a feminist analysis, it is perhaps more instructive to explore through this analysis the way in which they might be appropriated and controlled by *women* on their own behalf. Technologies *per se* are not necessarily dominating and exploitative of women: it is the way they are used and in what context that determines this. This allows the possibility that women are not, as Petchesky (1987) says, 'just passive victims of "male" reproductive technologies and the doctors who wield them'. In certain circumstances, perhaps associated with problems with pregnancy or infertility, reproductive techniques may *possibly* allow a woman to develop greater control over her reproductive capacity. And Petchesky notes, other groups without problems of fertility may too find some advantage in these techniques: 'The view that "reproductive engineering" is imposed on "women as a class" rather than being sought by them as a means towards greater "choice" obscures the particular reality, not only of women with fertility problems and losses, but also of other groups. For lesbians who utilise sperm banks and artificial insemination to achieve biological pregnancy without heterosexual sex, such technologies are a critical tool of reproductive freedom' (p.77). More generally, as Stanworth (1987) has argued, while a feminist challenge to reproductive technology needs to be made, this should not thereby lead to a position in which only 'natural' pregnancies should be advocated: this is to romanticise and mythologise female-centred images of motherhood. It is not evident, she says, 'what a "natural" relation to our fertility would look like, it is even less clear that it would be desirable: fertility undermined by poor nutrition or by gonorrhoea, unchecked by medical intervention; high birth rates, with population growth limited only by high infant and adult mortality; abstinence from intercourse for heterosexual people except when pregnancy was the immediately desirable result?' (p.34).

The critique of established science offered by feminists has been vital for raising historical, epistemological and theoretical questions about the direction and purpose of technologies (and not only those within the reproductive field). Like the 'alternative' and radical science movements considered earlier, feminist analysis provides a new way of exploring the processes through which scientific knowledge and technological practices become established, but like the other two, its insights have been only partly incorporated into mainstream sociology of science.

Conclusion

I have examined a number of what we can call 'populist' or alternative critiques of science and technology. Historically, there have been many pressure groups that have arisen in response to some perceived threat posed by scientific and technological developments, such as the Luddites, the anti-Darwinianists, and so on. We should not regard these movements as in some way irrational because they somehow 'prevent scientific progress': their very role is to challenge the simplistic idea that scientific development is ever progressive and admirable.

But, perhaps, the more recent critical movements have assumed a rather different character from those of the past. Some of the more successful challenges to established technological systems – such as has occurred with the growing political importance of environmentalism and 'deep green' politics (especially in Europe) – have achieved many of their goals because those involved have managed to reduce the institutional and cognitive *distance* between them and established science in both government and industry. As I noted earlier, Collins has pointed out that expertise is crucially dependent on being able to sustain its distance from lay people, thereby maintaining its apparent capacity for providing us all with 'certain' knowledge about the world. The more close one gets to expertise, the less certain it looks, as Nelkin and Latour also show through their respective examinations of controversy and laboratory life.

The groups we have considered in this chapter have become closer to science through their own experience – often as trained scientists – in the very areas they seek to challenge. Moreover, the established, orthodox 'republic of science' has become more complex and politicised, so that it is more difficult for elite agencies – whether in the professional and learned societies, research funding bodies, or via the institutional networks linking universities – to control both their own members and their status in the wider society. As the arena of science becomes more open, as new institutional structures promoted by both restructuring and commercialisation appear, and as technological systems become subject to state and international regulation through agencies such as the European Parliament, the distance between the scientific elites and wider society narrows and the capacity for these elites to sustain themselves in positions of high social status becomes more problematic. A very recent example of this happening has been the debate surrounding the 'human genome' project to map the genetic structure of the human body.

This is not to say that these changes herald the arrival of a more democratic and accountable science and technology: sometimes the effect may be precisely the opposite as the increasing number of agencies and actors involved make rational, democratic control more difficult to establish. The difficulty that has been associated with the harmonising of regulations concerning the release of genetically engineered organisms, in both the United States and Europe, is a case in point.

Nevertheless, all three movements explored in this chapter are based on the premise that the more one knows about science and technology the more possible it is to govern them in ways that serve wider social interests than has been the case thus far. As Wynne (1988) has suggested, 'technology should be seen as a large-scale social experiment, in which we are all involved. This in itself is no criticism, since it is unavoidable. However, the experimental character of technology is not part of public discourse, and its concealment only undermines a more constructive orientation which would be possible were this propery brought into the open' (p.163). These three movements have sought to bring technology into the public arena, to make it 'part of public discourse'. Many of those involved have drawn on ideas from philosophy, political theory, social ethics and scientific discourse itself: more recently, the contributions from social science and sociology in particular have grown, as empirical and theoretical work in sociology have demonstrated the *social* character of science and technology.

7 Conclusion

I have tried in this book to give the reader as wide a review of the principal areas of interest in the sociology of science today. I have been particularly keen to indicate how this work relates to important issues within the field of science policy, to show how this growing body of research can be regarded as more than simply an academic specialty within the discipline. In highlighting the major shifts in the sociology of science I have suggested that it has been able to explore more successfully and interestingly the new directions in which science and technology are moving today. These 'new directions' are not simply of a cognitive or conceptual nature – new fields of research for example, such as biotechnology – but also of an *institutional* character: in fact, both feed back on each other, cognitive developments shaping and being shaped by institutional developments. Biotechnology, for example, has opened new areas for investigation and new forms of investigation as the corporate/academic interface has created novel institutional structures within which the research has been pursued.

I argued that there are four main developments in science and technology today. Let me summarise them here. First, scientific labour inside the laboratory is typically being undertaken by interdisciplinary research teams or trans-laboratory networks whose research is less easily divided into the basic–applied dichotomy of the past: work in these teams has what we might call a 'co-valent' character, pulling in both 'basic' and 'applied' directions *simultaneously* but never in such a way as to completely separate the one from the other. It is very important to understand how these research teams are established, how mobility of staff between them occurs, and whether their research output is greater than smaller, more disaggregated styles of research more typical of the past (see P. Hoch, 1990).

Secondly, and as a direct result of the previous point, I have given considerable emphasis to the way in which science and technology have become less easy to distinguish *in practice* than in the past. Although science may be concerned with the understanding of general principles while technology seeks more specific applications thereof, scientific work is often today more interested in the development of techniques rather than general theories, though the techniques – such as polymerisation, or protein engineering – may have a general applicability. Moreover, the locational split conventionally assumed between scientists and technologists – the former in academe the latter in industry – cannot be sustained. Not only do the vast majority of scientists work in industrial (often militarily-related) contexts, industry will employ a large number of scientists to pursue long-term 'discovery' research in their search for the products that will bring in revenue in five to ten years' time.

Thirdly, the commercial exploitation of scientific knowledge has become necessary for the survival of firms in certain industrial sectors of the economy, those that are likely to provide the new manufacturing processes and products of the twenty-first century, such as bioscience, information technology and new material sciences. The exploitation of knowledge generated in-house by companies as well as externally in public sector research establishments requires effective access to information about state-of-the-art discoveries within particular areas of research. Companies have developed a wide range of strategies, both formal and informal, to achieve this objective.

Finally, science and technology have become subject to a growing promotion, monitoring and regulation by national and international agencies. More emphasis is given to mapping and evaluating research activity and increasingly so on behalf of the end-user of research rather than simply by peer review. State authorities have themselves sought to exploit their national science bases in order to maximise economic competitiveness. They have tried, more or less successfully, to do this through the twin mechanisms of restructuring and commercialisation (see Chapter 4). Scientific knowledge becomes more politicised and more contestable, less autonomous. More effort is made to involve the public in debate while the public itself makes more demands to be involved in policymaking, such as in British Public Inquiries, US 'Science Courts' or the Dutch 'PWT', the Foundation for Public Information on Science, Technology and the Humanities. However, as Collingridge and Reeve (1986) have argued, this makes science policymaking less easy than in the past since it is more difficult to produce a general consensus either about the 'evidence' itself or its implications. Expertise has become problematic.

These four developments in the institutional character of science and technology have posed a large number of research questions for sociologists of science. I argued that their ability to address let alone provide any tentative answers to them has depended on developments occurring within the field itself over the past twenty years. Chapters 1 and 2 provided an account of the shift from the early Mertonian position on science to one which explored the social interests, ethnographic features and discursive practices of scientists as part of a general adoption of a relativistic and constructionist approach in the sociology of science. The key differences between these schools were described, but I argued that it is possible that an emergent synthesis between them is becoming more apparent as we develop a picture of scientists as 'calculative' actors covering a terrain that goes beyond the laboratory to wider social arenas in which political and economic support must be enlisted.

The four trends highlighted above focus on the institutionalised character of science and technology, embedded in corporate, academic or state structures. There is, however, a wider sense in which we need to consider science and technology and that is as part of our general social culture, made up of taken-for-granted artefacts – such as cars or washing machines – which in turn depend on highly complex technological systems (see Hughes [1985]) in specific social and economic contexts. I argued in Chapter 3 that there is a strong tendency among science policymakers to 'black box' technology precisely because by doing so it appears easier to 'transfer', 'regulate', or 'assess'. Sociological research has, however, not merely opened but dismantled the black box, showing how our understanding and evaluation of technologies is socially constructed. This 'construction' process depends on continual negotiation, persuasion and competition among a whole range of social actors. To demonstrate this is not, however, to debunk or dismiss attempts by policymakers to develop and/or control science and technology: rather, I argued that sociological research into the nature of technological systems, expertise, or perceptions of technology held by the wider public, and so on, should help to provide a deeper awareness of the complexity of our technological culture and so develop a more sophisticated set of assumptions on which to base policy.

Preliminary findings from research in Britain on the public understanding of science by a number of sociologists demonstrate the variety of attitudes towards and experiences of science and technology among non-scientists as well as those working within science-related areas. For example, many industrial employees working in routine science-based jobs, such as trainee technicians, are unlikely either to know or want to know about the safety or even efficacy of their work as they assume this to be the responsibility of others more senior to them. We routinely assume that technologies embody the judgement of others who have preceded us or who are in authority, whether we are an engineer in a power station, a bacteriologist in a hospital laboratory, or a driver of a car. Our approach to technologies will depend on their place in our social relationships. From a policy perspective then, we should be trying to make more sense of people's perception of science and technology as experienced by them in their daily routines. As Schwarz and Thompson (1990) have argued, this means that the popularisation of science should not be based on some conventional image of what people 'need to know' about 'it', but more attuned to the everyday life and interests of the lay person. To achieve this it is likely that a plurality of agencies and media need to be developed: large, centralised agencies are unlikely to be sufficiently sensitive to the occupational, regional or institutional variety of perceptions towards and interests in science that people have.

As we saw in Chapter 4, however, the principal concern of the state in all OECD countries is to devise policies that will further the 'exploitation' of science and technology. In this context, it is more likely that educational programmes designed to raise people's understanding of science will convey messages about how science can benefit people without in fact addressing the sort of concerns that people actually have. The development of 'science shops' initially in the Netherlands and more recently in Northern Ireland are local attempts by natural and social scientists to rectify this problem by letting laypeople determine what needs to be discussed, explained and dealt with. These attempts at a more accessible and localised forum in which science and technology can be discussed are likely to remain few and far between as they lack wider political support from the state. If they are to grow they will probably only do so in those policymaking cultures that are both open and yet characterised by a degree of national planning: it is no surprise that science shops have had most success in the Netherlands.

It is, of course, important that information about science keeps pace not merely with the development of new ideas but also with the emergence of new institutional forms within which innovation occurs. Chapter 5 explored recent historical and sociological research on the changing patterns of corporate R&D and the new institutional relationships developing between industry and public sector science in universities and research institutes. These developments have created major concern and indeed conflict over such matters as who owns and control emergent knowledge, what the proper role for universities should be, how research agendas are being set and whether the science base is being skewed too much towards 'exploitable' rather than socially useful technology. As I illustrated in the discussion of biotechnology, these issues have only recently begun to be examined by sociologists. What information we have available does show that biotechnology has had a transformative effect on the relationships between academic and industrial science, as well as on the economy more generally in terms of the (dematerialised) character of productive activity it has allowed. Whether this effect can be deemed to be 'revolutionary' is, however, less certain. Nevertheless, the general point remains that any strategies that are designed to increase the public understanding of science should aim to increase awareness of not merely the content of science but also institutional changes new sciences are bringing for it is the latter rather than the former that is likely to have most immediate impact on people's lived experiences.

Future research in the sociology of science

While we have seen that there is a rich and varied stock of knowledge in the sociology of science there are a number of major gaps in our understanding that need to be filled. These gaps are both empirical and theoretical in nature.

There are five key substantive areas that future sociological research should consider exploring. First, as I suggested towards the end of Chapter 4, there has been relatively little research on the 'political economy' of the laboratory, that is, on the material and political relationships between those who work there and how these vary from one situation to another. Most empirical research has examined the role of those in positions of power – say the principal investigators on a research project – paying little or no attention to the large body of more junior research scientists and technicians that perform the routinised work-tasks of the lab. This contrasts strongly with sociological research elsewhere, in manufacturing or service industries, for example, where much is known about the labour process across a range of occupational groups enjoying greater or lesser degrees of power and authority in the organisation. As Yearley (1988) has commented, 'as industrial production becomes more technically sophisticated and as scientific work becomes increasingly machine-dependent, one would anticipate that science is becoming less distinct from other forms of work so that generalisations from the labour process literature could be tentatively applied to the laboratory' (p.16).

Secondly, as I have shown in Chapter 5, the commercial exploitation of science and technology is now a century-old phenomenon but taking on new dimensions as research-intensive industries such as biotechnology, pharmaceuticals, fibres and new materials, and information technology rely on innovation in order to remain competitive. Despite the importance of these research-intensive sectors there has been relatively little sociological research on the organisation and culture of the R&D laboratory inside companies. What work has been done suggests that company scientists are as elsewhere involved in negotiation over the status of their research (Vergragt, 1988) and have to engage in 'articulation' strategies (Fujimura, 1987) in order to translate ideas into commercially exploitable technologies. In doing so they are satisfying the different demands of different reference groups – other scientists, competitor scientists, product managers, accountants, divisional marketing staff and vice-presidents of research. At the same time, what these groups require not only differs between them but also changes over time within each. Potentially, then, this area is a very rich source of information on the way in which technological development is shaped and reshaped by the different interests –

cognitive, social and economic – of these groups. The ethnographic tradition in the sociology of science could be usefully combined with a more organisational approach, perhaps similar to that developed by Hull (1986) and others, extending thereby our sociological understanding of 'laboratory life' in contexts rather different from those found in academia. There is a pressing need for more research on scientific work within corporations that are located within the military–industrial sector.

Thirdly, stronger links need to be made between the contributions of feminists and those sociological schools examined in Chapters 1 and 2. They both show, though in different ways, how scientific authority is constructed and socially contingent. Feminists have been particularly interested in developing a theory of what a feminised science and technology might look like as well, which poses interesting questions for more mainstream sociology of science: what would, for example, 'industrially structured sciences with feminist values and goals' (Harding, 1986, p.302) look like? How would authority and expertise in science be differently produced, were these values to prevail? What direction would technological systems take if they were to be constructed according to these values rather than the 'masculinist' variety that has informed them thus far?

The fourth gap in current work lies between the sociology of science and the large body of work that has been produced – primarily by economists – covering the process of technological innovation and economic growth. Links here could be made through incorporating into the more economistic analysis sociological insights into the determination of technological 'success' that depends on more than market considerations and the production function. As I emphasised when discussing 'technology transfer' in Chapter 3, the boundaries of technologies and their 'innovative' status are not clearly defined: much depends on the negotiation and debate between more or less powerful interest groups. We need then to build these assumptions into analyses of the process and rate of innovation in order to determine whether there are any specific institutional or more general cultural features which constrain or promote what can be accepted as 'innovative' development.

Finally, there is at present a desire on the part of both sociologists and science policymakers to develop stronger links between them.

There is, however, still some uncertainty as to what this might involve: some balance is needed between the demands of the policymaker for 'relevance' and the academic research interests of the field. The latter must not be compromised simply to serve the former. This means that most often, sociology can act to help shape or enlighten the policymaking process rather than being directly involved in it. I suggested that policymaking is far from a tidy, rational affair anyway, since it is itself informed by all sorts of value-judgements held by different interested parties operating in arenas that may be more or less open and more or less centrally planned. It would then be somewhat naive for sociologists to think that their research findings can be introduced into this process without many of them being transformed or lost along the way. At the same time, I believe that it is unrealistic for policymakers to demand that sociology should have some policy pay-off since this presupposes that there is a well-oiled policymaking machine into which sociological information could be fed. Nevertheless, as Cozzens (1986) argued, this should not prevent both groups from trying to engage in a dialogue with each other about science and technological development and its impact on a wider citizenry. But sociologists should not become yet another set of 'high priests' in the rituals of policymaking. They should encourage a more open and accountable promotion and monitoring of science and technology in order to make both more accessible to those most likely to be affected by them.

Conclusion

A piece of sociology of science fiction

. . . Some years earlier, the Discovery Dome had been replaced by the 'Black Box'. Inside the massive monolithic structure was a maze of routes that led to the Interpretation Windows. Visitors to the museum could construct different images of technoscience using their interprecorders at any Window they cared.

Jay was enjoying reconstructing the story of the human genome project. Suddenly, a loud, violent series of blows shattered her image. Above her two men were hammering at one of the Windows in a frenzied attempt to seal it shut. They wore emblems on their tunics that bore the logo SIR of the radically conservative 'Science Is Right' campaign. Until recently, SIR had been regarded as an unimportant sect of Certaintists trying to restore the power of science. But since their new leader Reppop had appeared on the scene, they had grown more militant and daring in their raids on the Museums of Reflexive Science.

Jay thought for one second that she might be imagining all this, that her interprecorder was malfunctioning. But a stray bolt from the polychromatic laser hammer told her otherwise. The next thing she knew . . .

So, it's not Isaac Asimov, I admit. But a possible scenario? No: I doubt that our science and technology culture will ever become so institutionally reflexive as this, or that a group like 'SIR' would ever be formed. 'Technoscience' is located in and reproduced by powerful economic and political sectors while the push for greater democratic control over both is limited though not insignificant.

It is more likely that the four directions in which science and technology are moving will bring a new institutional strength to both. All four encourage the more effective exploitation of science and technology through state, trans-state and corporate activities. The institutional structures that are being put in place to monitor research and development and their wider environmental effect may generate more debate about science and so limit the capacity of scientists themselves to determine how to handle innovation (as they were able to do, for example, in the mid-1970s with genetic engineering [Wright, 1986]. However, the national and inter-national agencies that have been established to monitor science have sought to establish codes of conduct and good practice within the *status quo* rather than entirely new forms of accountability or orientation.

So my story about the future science museum will remain fictitious. But as they say, truth is stranger than fiction: only the other day I was in the British Science Museum contemplating the merits of an exhibit recounting the conventional history about the development of the bicycle when I was astounded to overhear a group of people nearby talking about the very different tech-nological routes which the development of the machine might have taken. It was then that I saw the 'Sociology Conference' badges . . .

References

1 Introduction

Advisory Council for Applied Research and Development (1986) *Exploitable Areas of Science* (London: HMSO).

Aronowitz, S. (1989) *Science as Power* (London: Macmillan).

Bourdieu P. (1975) 'The Specificity of the Scientific Field and the Social Conditions of the Progress of Reason', *Social Science Information*, Vol. 14, pp.19–47.

Braverman, H. (1974), *Labour and Monopoly Capital* (New York: Monthly Review).

Carson, R. (1965) *Silent Spring* (Harmondsworth: Penguin).

Ciccotti, G. (1976) 'The Production of Science in Advanced Capitalist Societies', in H. and S. Rose (eds), *The Political Economy of Science* (London: Macmillan).

Collins, H.and Pinch, T. (1979) 'The Construction of the Paranormal: Nothing Unscientific is Happening', in R. G. A. Dolby (ed.), *On the Margins of Science* (Keele: University of Keele).

Collins, H. (1975) 'The Seven Sexes: A Study in the Sociology of a Phenomenon, or the Replication of Experiments in Physics', *Sociology*, Vol. 9 (2), pp.205–24.

Collins, H. (1981) 'Stages in the Empirical Programme of Relativism', *Social Studies of Science*, Vol. 11 (1), pp.3–10.

Collins, H (1985) *Changing Order* (London: Sage).

Dennis, M. (1987) 'Accounting for Research: New Horizons of Corporate Laboratories and the Social History of American Science', *Social Studies of Science*, Vol. 17, pp.179–518.

Dolby, R. G. A. (1972) 'The Sociology of Knowledge in Natural Science' in B. Barnes (ed.), *Sociology of Science* (Harmondsworth: Penguin).

Epstein, S. (1979) *Politics of Cancer* (New York: Doubleday).

Freeman, C. (1987) *Technology Policy and Economic Performance* (London: Frances Pinter).

Gillespie, B. *et al.* (1979) 'Carcinogenic Risk Assessment in the United States and Great Britain', *Social Studies of Science*, Vol. 9, pp. 265–302

Gorz, A. (ed.) (1976) *The Division of Labour: The Labour Process and Class Struggle in Modern Capitalism* (London: Harvester Press).

Hessen, B. (1931) 'The Social and Economic Roots of Newton's *Principia*', in N. I. Bukharin *et al.* (eds), *Science at the Crossroads* (second edition, 1971) (London: Frank Cass).

Hounshell, D. (1988) *Science and Corporate Strategy: Dupont R&D 1902–1980* (Cambridge: Cambridge University Press).

Kuhn, T. (1970) *The Structure of Scientific Revolutions* (second edition) (Chicago: University of Chicago Press).

Levidow, L. and B. Young (eds) (1982) *Science, Technology and the Labour Process* (Vol. 2) (London: Humanities Press).

Merton, R. K. (1949) *Social Theory and Social Structure* (New York: Free Press).

Mulkay, M. J. (1976a) 'Norms and Ideology in Science', *Social Science Information*, Vol. 15, pp.637–56.

Mulkay, M. J. (1976b) 'The Mediating Role of the Scientific Elite', *Social Studies of Science*, Vol. 6, pp.445–70.

Mulkay, M. J. (1979) *Science and the Sociology of Knowledge* (London: George Allen & Unwin).

Mulkay, M. J. (1984) *Opening Pandora's Box* (Cambridge: Cambridge University Press).

Mulkay, M. J. and D. O. Edge (1976) *Astronomy Transformed* (New York: Wiley Interscience).

Navarro, V. (1986) *Crisis, Health and Medicine: a social critique* (London: Tavistock).

Nelkin, D. (1979) *Controversy: politics of technical decisions* (New York: Sage).

Popper, K. (1972) *Objective Knowledge: an evolutionary approach* (Oxford: Oxford University Press).

Price, D. (1984) *Little Science, Big Science* (second edition) (New York: Columbia University Press).

Rose, H. and S. (eds) (1976) *The Political Economy of Science* (London: Macmillan).

Science and Engineering Research Council (1987), 'Switched On Cells and Anti-Sense Genes', *Biobulletin*, Vol. 5, no. 1, October, p.6.

Storer, N. W. (1966) *The Social System of Science* (London: Holt, Rinehart and Winston).

Webster, A. (1979) 'Scientific Controversy and Socio-Cognitive Metonymy' in R. G. A. Dolby (op. cit.).

Webster, A. (1981) *Structural and Interpretive Resources in Science* PhD thesis (unpublished), University of York.

Woesler, C. (1976) 'Opposition between professionalisation and political practice in West German Sociology', *Social Science Information*, Vol. 15, pp.663–87.

Wright, P (1979) 'A Study in the Legitimisation of Knowledge: The "Success" of Medicine and the "Failure" of Astrology', in R. G. A. Dolby (op. cit.).

Wynne, B. (1979) 'Between Orthodoxy and Oblivion: The Normalisation of Deviancy in Science', in R. Wallis (ed.), *On the Margins of Science* (Keele: University of Keele).

2 Sociology of Science and Technology

Barnes, B. and D. MacKenzie (1979) 'On the Role of Interests in Scientific Change', in R. Wallis (ed.) (op. cit.).

Bloor, D. (1976) *Knowledge and Social Imagery* (London: Routledge and Kegan Paul).

Chubin, D. and S. Restivo (1983) 'The "Mooting" of Science Studies: Research Programmes and Science Policy', in K. Knorr-Cetina and M. J.Mulkay (eds) *Science Observed* (London: Sage).

Collins, H. (1981) *Knowledge and Controversy: Studies of Modern Natural Science*, special edition of *Social Studies of Science*, Vol. 11 (1).

Collins, H. (1983) 'An Empirical Relativist Programme in the Sociology of Scientific Knowledge', in K. Knorr-Cetina and M. J. Mulkay (loc. cit.).

Collins, H. (1985) *Changing Order* (London: Sage).

Corea, G. et al. (1985) *Man-made Women* (London: Hutchinson).

Cowan, R. S. (1979) 'Virginia Dare to Virginia Slims: Women and Technology in American Life', *Technology and Culture*, Vol. 20, pp.51–63.

Dean, J. (1979) 'Controversy over classification', in B. Barnes and S. Shapin (eds), *Natural Order* (London: Sage).

Fuhrman and Oeler (1986) 'Discourse Analysis', *Social Studies of Science*, Vol. 16 (2), pp.300–15.

Garfinkel, H. (1967) *Studies in Ethnomethodology* (Prentice-Hall: Englewood Cliffs: New Jersey).

Gieryn, T. (1982) 'Relativist/Constructivist Programmes in the Sociology of Science: Redundance and Retreat', *Social Studies of Science*, Vol. 12, pp.279–97.

Gilbert, G. N. and M. J. Mulkay (1984) *Opening Pandora's box* (Cambridge: Cambridge University Press).

Goodfield, J. (1981) *An Imagined World* (New York: Harper & Row).

Halfpenny, P. (1988) 'Talking of Talking, Writing of Writing: some reflections on Gilbert and Mulkay's Discourse Analysis', *Social Studies of Science*, Vol. 18 (1), pp.169–82.

Halfpenny, P. (1989) 'Reply to Potter and McKinlay', *Social Studies of Science*, Vol. 19 (1), pp.146–51.

Hughes, T. (1983) *Networks of Power: Electrification in Western Society 1880–1930* (London: Johns Hopkins University Press).

Knorr-Cetina, K. and M. J. Mulkay (1983) *Science Observed* (London: Sage).

Latour, B. (1987) *Science in Action* (Milton Keynes: Open University Press).

Latour, B. and S. Woolgar (1979) *Laboratory Life: the construction of scientific facts* (Princeton: Princeton University Press).

MacKenzie, D. (1987) 'Missile Accuracy: A Case-Study in the Social Process of Technological Change,' in W. Bijker *et al.*, *The Social Construction of Technological Systems* (London: MIT).

Mulkay, M. J. *et al.* (1983) 'Why an Analysis of Scientific Discourse is Needed', in K. Knorr-Cetina and M. J. Mulkay (eds) (op. cit).

Parssinen, T. M. (1979) 'Professional Deviants in Scientific Change', in R. Wallis (ed.) (op. cit.).

Pinch, T. J. (1981) 'The Sun-Set: the Presentation of Certainty in Scientific Life', *Social Studies of Science*, Vol. 11 (1), pp.131–58.

Pinch, T. and W. Bijker (1987) 'The Social Construction of Facts and Artefacts', in Bijker, W. *et al.* (eds), *The Social Construction of Technological Systems* (London: MIT Press).

Rose, H. and J. Hanmer (1976) 'Women's Liberation: Reproduction and the Technological Fix', in H. and S. Rose (eds), *The Political Economy of Science* (London: Macmillan).

Shapin, S. (1979) 'The Politics of Observation: Cerebral Anatomy and Social Interests in the Edinburgh Phrenology Disputes', in R. Wallis (ed.) (op. cit.).

Shapin, S. (1984) 'Talking History: Reflections on DA', *Isis*, Vol. 75, pp.125–8.

Shapin, S. (1988) 'Following Scientists Around', *Social Studies of Science*, Vol. 18, pp.533–50.

Stanworth, M. (ed.) (1987) *Reproductive Technologies* (Cambridge: Polity).

Travis, D. *et al.* (1986) *The Politics of Uncertainty* (London: Routledge & Kegan Paul).

Woolgar, S. (1976) 'Writing an intellectual history of scientific development: the use of discovery accounts', *Social Studies of Science*, Vol. 6, pp.395–422.

Woolgar, S. (1981) 'Interests and Explanation in the Social Study of Science', *Social Studies of Science*, Vol. 11, pp.365–94.

Woolgar, S. (1988a) *Science: the Very Idea* (London: Tavistock).

Woolgar, S. (ed.) (1988b) *Knowledge and Reflexivity: new frontiers in the sociology of knowledge* (London: Sage).

Zenzen, M. and Restivo, S. (1982) 'The Mysterious Morphology of Immiscible Fluids: A Study of Scientific Practice', *Social Science Information*, Vol. 23, pp.447–73.

3 Sociology of Science Policy: Opening and Managing the 'Black Box'

Advisory Council for Applied Research and Development (1986) *Exploitable Areas of Science* (London: HMSO).

Alter, P. (1987) *The Reluctant Patron: Science and the State in Britain 1850–1920* (London: Berg).

Annerstedt, J. and Jamison, A. (1988) *From Research Policy to Social Intelligence* (London: Macmillan).

Barnes, A. D. (1987) (ed.) *Technology Transfer: A European Perspective* (Sheffield: University of Sheffield).

Bohme, G. (1983) 'Introduction: the social determinants of knowledge', in W. Schafer (ed.), *Finalisation in Science: The Social Orientation of Scientific Progress* (Dordrecht: Reidel).

Brickman, R. *et al.* (1986) *Controlling Chemicals: the politics of regulation in Europe and the United States* (Ithaca: Cornell University Press).

Butcher, J. (1987) 'Opening Address', in G. Ashworth and A. Thornton (eds), *Technology Transfer and Innovation* (London: Taylor Graham).

Cabinet Office (1986) *Annual Review of Government Funded R&D* (London: HMSO).

Clark, N. (1985) *The Political Economy of Science and Technology* (Oxford: Blackwell).

Collins, H. (1982) 'The TEA Set: Tacit Knowledge and Scientific Networks', in B. Barnes and D. Edge (eds), *Science in Context: Readings in the Sociology of Science* (Milton Keynes: Open University).

Collins, H. (1985) *Changing Order* (London: Sage).

Collingridge, D. and C. Reeve (1986) *Science Speaks to Power: The Role of Experts in Policy Making* (London: Frances Pinter).

Conroy, R. (1988) 'Technology and Economic Development', in R. Benewick and P. Wingrove (eds), *Reforming the Revolution China in Transition* (London: Macmillan).

Coombs, R. et al. (1987) *Economics and Technical Change* (London: Macmillan).

Cozzens, S. E. (1986) 'Funding and Knowledge Growth', *Social Studies of Science*, Vol. 16, pp.9–21.

Freeman, C. (1974) *The Economics of Industrial Innovation* (Harmondsworth: Penguin).

Freeman C. (1988) 'Qualitative and Quantitative Factors in National Technology Policies', in J. Annerstedt and A. Jamison (eds), *From Research Policy to Social Intelligence* (London: Macmillan).

Gummett, P. (1986) 'How to Influence Government Policy', *Science and Public Policy*, Feb., pp.60–61.

Guy, K. et al. (1987) *Evaluation of the Alvey Programme* (London: HMSO).

Hawkes, N. (1986) *The Worst Accident in the World: Chernobyl, the end of the nuclear dream* (London: Heinemann).

HMSO (1914–16) *Scheme for the Organisation and Development of Scientific and Industrial Research*, Cd 8005, 1.

HMSO (1979) *The Control of Radioactive Wastes: a review of Cmnd 884* (London: HMSO).

Irvine, J. and B. Martin (1984) *Foresight in Science: Picking the Winners* (London: Frances Pinter).

Irwin, A. (1985) *Risk and the Control of Technology* (Manchester: Manchester University Press).

Irwin, A. and P. Vergragt (1989) 'Rethinking the relationship between environmental regulation and industrial innovation: the social negotiation of technical change', *Technology Analysis and Strategic Management*, Vol. 1 (1), pp.57–70.

Ives, J. (1986) *Transnational Corporations and Environmental Control* (London: Routledge & Kegan Paul).

Jasanoff, S. S. (1987) 'Contested Boundaries in Policy Related Science', *Social Studies of Science*, Vol. 17, pp.195–230.

Kamin, L. J. (1974) *The Science and Politics of IQ* (Maryland: Erlbaum).

Lindblom, C. (1980) *The Policy Making Process* (second edition) (Englewood Cliffs, NJ: Prentice-Hall).

Martin, B. and J. Irvine (1985) 'Evaluating the Evaluators', *Social Studies of Science*, Vol. 15, pp.558–75.

National Science Foundation (1989) *Science Literature Indicators* (Washington: NSF) (annually).

Nelkin, D. (1979) *Controversy: politics of technical decisions* (New York: Sage).

Ronayne, J. (1984) *Science in Government* (London: Edward Arnold).

Rose, S. (1976) 'Scientific Racism and Ideology: The IQ Racket from Galton to Jensen', in H. & S. Rose (eds), *The Political Economy of Science* (London: Macmillan).

Rothwell, R. (1984) 'Difficulties of National Innovation Policies,' in M. Gibbons *et al.* (eds), *Science and Technology Policy in the 1980s and Beyond* (Harlow: Longman).

Schafer, W. (1983) 'The finalisation debate', in W. Schafer (1983) (op. cit.)

Tierney, S. F. (1979) 'The Nuclear Waste Disposal Controversy', in D. Nelkin (op. cit.) pp.91–110

Vogel, D. (1986) *National Styles of Regulation: Environmental Policy in Great Britain and the United States* (Ithaca: Cornell University Press).

Warnock, M. (1985) *A Question of Life: The Warnock Report on Human Fertilisation and Embryology* (Oxford: Blackwell).

Wynne, B. (1982) *Rationality and Ritual: The Windscale Inquiry and Nuclear Decisions in Britain* (Chalfont St Giles: British Society for the History of Science).

Wynne, B. 'Not in My Backyard', *The Times Higher Education Supplement*, 25 November 1988, pp.20–21.

Yearley, S. (1987) *Science, Technology and Social Change* (London: Unwin Hyman).

4   Exploiting Science and Technology (I)

ACOST (1989) *Defence R&D: A National Resource* (London: HMSO)

Arnold, E. and K. Guy (1986) *Parallel Convergence* (London: Frances Pinter).

Brickman, R. *et al.* (1985) *Controlling Chemicals: the politics of regulation in Europe and the United States* (Ithaca: Cornell University Press).

Bullock, M. (1983) *Academic Enterprise, Industrial Innovation, and the Development of High Technology Financing in the USA* (London: Brand Brothers).

Cabinet Office (1989) *Annual Review of Government Funded Research and Development* (London: HMSO).

Callon, M. (1979) *Proxan: a visual display technique for scientific and technical problem networks* (Paris: OECD).

Clark, N. (1985) *The Political Economy of Science and Technology* (Oxford: Blackwell).

Clarke, A. *et al.* (1984) 'University Autonomy and Public Policies', *Higher Education*, Vol. 13, pp.23–48.

Cowan, T. and F. Buttel (1987) 'Subnational Corporatist Policy-making: the organisation of state and regional high-technology development', mimeo, Dept of Sociology, Cornell University, Ithaca.

Currie, J. (1987) *Science Parks in Britain: their role for the late 1980s* (Cardiff: CSPE Economic).

The Economist (1985) 'Planting science parks in the UK', 16 March, pp.88–9.

Epstein, S. (1979) *The Politics of Cancer* (London: Fontana).

Etzkowitz, H. (1989) 'Entrepreneurial Science in the Academy: A Case of the Transformation of Norms', *Social Problems*, Vol. 36 (1), pp.14–27.

Faulkner, W. (1986) 'Linkage Between Industry and Academic Research: The Case of Biotechnology Research in the Pharmaceutical Industry', PhD thesis (unpublished) SPRU, University of Sussex.

Freeman, C. (1987) *Technology Policy and Economic Performance* (London: Frances Pinter).

Gibbons, M. *et al.* (1985) *Science as a Commodity* (Harlow: Longman).

Gillespie, B. *et al.* (1982) 'Carcinogenic risk assessment in the USA and UK: the case of Aldrin/Dieldrin', in B. Barnes and D. Edge, *Science in Context* (Milton Keynes: Open University Press).

Gummett, P. and Walker, W. (1989) 'European Defence Procurement and Industrial Capabilities', *Technology Analysis and Strategic Management*, Vol. 1 (2), pp.191–204.

Guy, K. (1987) *Evaluation of the Alvey Programme* (London: HMSO).

Ince, M. (1983) *The Politics of British Science* (Brighton: Wheatsheaf).

Jasanoff, S. (1989) 'Public participation in science policy', *Chemistry in Britain*, April, pp.368–70.

Kloppenberg, J. (1989) *First the Seed* (London: Yale University Press).

Lowe, J. (1985) 'Science Parks in the UK', *Lloyds Bank Review*, April, pp.31–42.

Marvin, S. J. (1987) *Local Authority Technology Development Policies and Initiatives: An Overview* (Milton Keynes: Open University Press).

Massey, D. (1985) 'Whose New Technology?' in M. Castells (ed.), *High Technology, Space and Society* (London: Sage).

Mathias, P. (1986) *Report of the Working Party on the Private Sector Funding of Scientific Research* (London: HMSO).

Merrison Report (1982) *Report of a Joint Working Party on the Support of University Scientific Research*, Advisory Board for the Research Councils (London: HMSO).

Muir-Wood, R. (1983) *Improving Research Links Between Higher Education and Industry* (London: HMSO).

Mulkay, M. J. (1976) 'Norms and Ideology in Science', *Social Science Information*, Vol. 15, pp. 637–56.

Mulkay, M. J. (1979) *Science and the Sociology of Knowledge* (London: George Allen & Unwin).

Nelkin, D. (1984) *Science as Intellectual Property* (London: Macmillan).

OECD (1984) *Industry and University, New Forms of Cooperation and Communication* (Paris: OECD).

Pavitt, K. (1983) *Patterns of Technical Change – Evidence, Theory and Policy Implications* (Sussex: SPRU).

Powers, D. R. *et al.* (1988) *Higher Education in Partnership with Industry* (London: Jossey-Bass).

Read, N. (1989) 'The near market concept applied to UK agricultural research', *Science and Public Policy*, Vol. 16 (4), pp.233–8.

Reece, C. (1986) *Exploitable Areas of Science*, ACARD (London: HMSO).

Robbins, K. and F. Webster (1988) 'Athens without slaves . . . or slaves without Athens', *Science As Culture*, Vol. 3, pp.7–53.

Rothman, H. (1984) 'Science Mapping for Strategic Planning', in M. Gibbons *et al.*, *Science and Technology Policy in the 1980s and Beyond* (Harlow: Longman).

Rothschild, E. (1985) 'Science and Society – A Changing Relationship', in M. Gibbons and B. Wittrock (eds) *Science as a Commodity* (Harlow: Longman).

Rothschild Lord (1971) 'The Organisation and Management of Government R&D', *A Framework for Government R&D* (The Rothschild Report) Green Paper Cmnd 4814 (London: HMSO).

Rothwell, R. and M. Dodgson (1989) 'Technology-based small and medium sized firms in Europe', *Science and Public Policy*, Vol. 16 (1), pp.9–18.

Russell, M. G. and D. J. Moss (1989) 'Science Parks and Economic Development', *Interdisciplinary Science Reviews*, Vol. 14 (1).

Scott, J. (1982) *The Upper Classes* (London: Macmillan).

Scott, P. (1984) *The Crisis of the University* (Beckenham: Croom Helm).

Segal, Quince (1985) *The Cambridge Phenomenon* (Cambridge: Segal Quince & Partners).

Shapin, S. (1988) 'Following Scientists Around', *Social Studies of Science*, Vol. 18, pp.533–50.

Simmie, J. and N. James (1986) 'The money map of defence', *New Society*, January 31, pp.179–80.

Smith, D. C. *et al.* (1986) 'National Performance in Basic Research', *Nature*, Vol. 323, 23 October, pp.681–6.

Swinbanks, D. (1989) 'Who should decide policy?' *Nature*, Vol. 338 (16), March, p.189.

Vogel, D. (1986) *National Styles of Regulation: environmental policy in Great Britain and the United States* (Ithaca: Cornell University Press).

Webster, A. and J. Constable, (1989) 'Emergent Research Alliances', *Industry and Higher Education*, December.

Webster, A. and J. Constable (1990) 'Strategic Alliances', *Industry and Higher Education*, December.

Webster, A. J. (1990) 'The Incorporation of Biotechnology into Plant Breeding in Cambridge', in I. Varcoe *et al.* (eds), *Deciphering Science and Technology* (London: Macmillan).

Weiner, M. (1981) *English Culture and the Decline of the Industrial Spirit* (Cambridge: Cambridge University Press).

Wield, D. (1986) *The Politics of Technological Innovation* TPG Occasional Paper 10 (Milton Keynes: Open University Press).

Wright, S. (1986) 'Recombinant DNA Technology and its Social Transformation, 1972–1982', *Osiris*, 2nd series, Vol. 12, pp.303–60.

Yearley, S. (1988) *The Occupational Culture, Organisation and Exploitability of Scientific Work*, SPSG Concept Paper no. 6.

Yoxen, E. (1986) *Unnatural Selection? Coming to terms with the new genetics* (London: Heinemann).

Ziman, J. (1987) 'The problems of "Problem Choice"', *Minerva*, Vol. 25 (1–2), pp.92–106.

Ziman, J. (1988) *Science in a Steady State*, SPSG Concept Paper no. 1 (London).

Ziman, J. (1989) *Restructuring Academic Science: A New Framework for UK Policy*, SPSG Concept Paper no. 8 (London).

5 Exploiting Science and Technology (II)

Bartels, D. and R. Johnston (1984) 'The Sociology of Goal-Directed Science: Recombinant DNA Research', *Metascience*, Vol. 1, pp.37–45.

Braverman, H. (1974) *Labour and Monopoly Capital* (New York: Monthly Review).

Burns, T. and G. M. Stalker (1961) *The Management of Innovation* (London: Tavistock).

Buttel, F. (1988) 'Are High-technologies Epoch-Making Technologies? The Case of Biotechnology', mimeo, Dept of Rural Sociology, Cornell University.

Clark, J. *et al.* (1988) *The Process of Technological Change: new technology and social choice in the workplace* (Cambridge: Cambridge University Press).

Coombs, R. *et al.* (1987) *Economics and Technical Change* (London: Macmillan).

Cowan, R. S. (1987) 'The Consumption Junction: a proposal for research strategies in the sociology of technology', in W. E. Bijker et al. (eds), *The Social Construction of Technological Systems* (Cambridge, Mass: MIT Press).

Crespi, R. (1989) 'Claims on tissue plasminogen activator', *Nature*, Vol. 337, 26 Jan, pp.317–8.

Cressey, P. and J. McInnes (1980) 'Voting for Ford: Industrial Democracy and the Control of Labour', *Capital and Class*, Vol. 11, pp.5–33.

Dennis, M. A. (1987) 'Accounting for Research: New Histories of Corporate Laboratories and the Social History of American Science', *Social Studies of Science*, Vol. 17, pp.479–518.

European Commission (1983) *Biosociety*, FAST/ACPM/79/14–3E (Brussels).

European Federation of Biotechnology (1988) 'Biotechnology in Europe' *Trends in Biotechnology*, Vol. 6, Sept., pp.207–8.

Fairtclough, G. (1985) 'Make me a molecule: the technology of recombinant DNA', *Proceedings of the Royal Institution*, Vol. 157 (Middlesex: Science Reviews Limited).

Freeman, C. (1977) 'Economics of Research and Development', in I. Spiegel-Rosing and D. Price (eds), *Science, Technology and Society* (London: Sage).

Freeman, C. (1982) *The Economics of Industrial Innovation* (London: Frances Pinter).

Fujimura, J. (1987) 'Constructing Do-able Problems in Cancer Research: Articulating Alignment', *Social Studies of Science*, Vol. 17, pp.257–93.

Hales, M. (1986) 'Management Science and the Second Industrial Revolution', in L. Levidow (ed.), *Radical Science* (New Jersey: Free Association Books).

Hillman, H. (1972) *Certainty and Uncertainty in Biochemical Techniques* (Henley: Surrey University Press).

Hounshell, D. and J. K. Smith (1988) *Science and Corporate Strategy: Du Pont R&D 1902–1980* (Cambridge: Cambridge University Press).

Hughes, T. (1987) 'The Evolution of Large Technological Systems', in W. E. Bijker *et al.* (eds), The Social Construction of Technological Systems (London: MIT Press).

Hull, F. (1988) 'Inventions from R&D: Organisational Designs for Efficient Research Performance', *Sociology*, Vol. 22 (3), pp.393–415.

Kloppenberg, J. (1988) *First the Seed: the political economy of plant biotechnology* (Cambridge: Cambridge University Press).

Latour, B. (1987) *Science in Action* (Milton Keynes: Open University Press).

Mowery, D. C. and N. Rosenberg (1989) *Technology and the Pursuit of Economic Growth* (New York: Cambridge University Press).

Moser, A. (1988) 'Biotechnology – quo vadis? Scientia sine conscientia?', *Trends in Biotechnology*, Vol. 6, Sept., pp.207–8.

Newman, A. (1990) 'Experiments with automated lab staff', *Chemistry and Industry*, 5 Feb., pp.56–8.

Noble, D. F. (1985) 'Social Choice in Machine Design: the case of automatically controlled machine tools', in Mackenzie and J. Wajcman (eds), *The Social Control of Technology* (Milton Keynes: Open University Press).

Office Of Technology Assessment (OTA) (1988) *US Investment in Biotechnology* (Washington, DC: DTA Government Printing Office).

Palloix, C. (1976) 'The Labour Process from Fordism to Neo-Fordism', in *The Labour Process and Class Strategies* (Stage 1), pp.46–67.

Panem, S. (1984) *The Interferon Crusade* (Washington, DC: The Brookings Institution).

Primrose, S. B. (1988) *Modern Biotechnology* (Oxford: Blackwell).

Reich, L. (1985) *The Making of American Industrial Research: Science and Business at General Electric and Bell 1876–1926* (Cambridge: Cambridge University Press).

Sargeant, K. (1984) 'Biotechnology, connectedness and dematerialisation: the strategic challenges to Europe and the Community response', Paper presented to Biotechnology '84 (Dublin: Royal Irish Academy).

Sharp, M. (1986) *The New Biotechnology*, Sussex European Papers, no. 15 (Brighton: SPRU).

Sohn-Rethel, A. (1978) *Intellectual and Manual Labour* (London: Macmillan).

Stent, G. (1968) 'That was the molecular biology that was', *Science*, Vol. 160, pp.390–95.

Stoneman, P. (1988) *The Economic Analysis of Technology Policy* (Oxford: Oxford University Press).

Swann, J. P. (1988) *American Scientists and the Pharmaceutical Industry* (Baltimore: Johns Hopkins Press).

Tait, J. (1989) 'Proactive Risk Regulation for Biotechnology Products', Centre for Technology Strategy (Milton Keynes: Open University), mimeo.

Trevan, M. *et al.* (1988) *Biotechnology: the biological principles* (Milton Keynes: Open University Press).

Vergragt, P. (1988) 'The Social Shaping of Industrial Innovations', *Social Studies of Science*, Vol. 18, pp. 483–513.

Watson, J. D. (1968) *The Double Helix* (Harmondsworth: Penguin).

Wright, S. (1986) 'Recombinant DNA and its social transformation 1972–82,' *Osiris*, 2nd series, Vol. 12, pp.303–60.

Yoxen, E. (1983) *The Gene Business* (London: Pan Books).

6   Controlling Science and Technology: Popular and Radical Alternatives

Anderson, G. C. (1990) 'Congress cracks down', *Nature*, Vol. 343, 15 Feb. p.580.

Beier, R. (ed.) (1986) *Feminist Appoaches To Science* (New York: Pergamon Press).

Bernal, J. D. (1939) *Social Function of Science* (London: Routledge).

Boggs, C. (1986) *Social Movements and Political Power* (Philadelphia: Temple University Press).

Clark, N. (1985) *The Political Economy of Science and Technology* (Oxford: Blackwell).

Cole, J. (1979) *Fair Science: Women in the Scientific Community* (New York: Free Press).

Collins, H. M. (1987) 'Certainty and the Public Understanding of Science: Science on Television', *Social Studies of Science*, Vol. 17, pp.689–713.

Corea, G. *et al.* (1985) Man-made Women (London: Hutchinson).

Cromwell, G. (1989) 'Government policy and alternative strategies for appropriate technology choice', *Science and Public Policy*, Vol. 16 (4), pp.202–10.

Delamont, S. (1987) 'Three Blind Spots?' *Social Studies of Science*, Vol. 17, pp.163–70.

Dickson, D. (1984) *The New Politics of Science* (New York: Pantheon).

Einstein, A. (1946) 'Why Socialism?', *Monthly Review* (Hebrew University).

Ellman, L. (1990) *Guardian*, 5 Feb., p.20.

Elzinga, A. (1988) 'Bernalism, Comintern and the Sciences of Science: Central Science Movements Then and Now', in J. Annerstadt and A. Jamison (eds), *From Research Policy to Social Intelligence* (London: Macmillan).

French, R. D. (1975) *Antivivisectionism and Medical Science in Victorian Society* (Princeton: Princeton University Press).

Griffiths, D. (1985) 'The exclusion of women from technology', in W. Faulkner and E. Arnold (eds), *Smothered by Invention* (London: Pluto Press).

Ince, M. (1986) *The Politics of British Science* (Brighton: Wheatsheaf).

Inglehart, R. (1987) *The Silent Revolution: Changing Values and Political Styles Among Western Publics* (Princeton: Princeton University Press).

Jones, G. *et al.* (1978) *The Presentation of Science by the Media* (Primary Communications Research Centre: University of Leicester).

Jordanova, L. (1989) *Sexual Visions: images of gender in science and medicine between the eighteenth and twentieth centuries* (Brighton: Harvester Wheatsheaf).

Keller, E. F. (1985) *Reflections on Gender and Science* (London: Yale University Press).

Kuznick, P. J. (1987) *Beyond the Laboratory: scientists as political activists in 1930s America* (Chicago: University of Chicago Press).

Lewis, G. (1986) 'Regulatory Mechanisms and the Control of Medicinal Drugs – the case of DepoProvera', *Science, Technology Society Association Newsletter*, no 24, Spring, pp.43–7.

Lowe, P. and J. Goyder (1983) *Environmental Groups in Politics* (London: Allen & Unwin).

Lukes, S. (1979) *Power: A Radical View* (London: Macmillan).

McRobie, G. (1980) *Small Is Possible* (New York: Harper and Row).

Nelkin, D. (1984) *Controversy* (second edition) (London: Sage).

Newby, H. (1987) *Country Life* (London: Weidenfeld & Nicolson).

Papadakis, E. (1988) 'Social Movements, Self-Limiting Radicalism and the Green Party in West Germany', *Sociology*, Vol. 22 (3), pp.433–54.

Petchesky, R. P. (1987) 'Foetal Images: the Power of Visual Culture in the Politics of reproduction', in M. Stanworth (ed) *Reproductive Technologies* (Cambridge: Polity Press).

Ravetz, J. R. (1982) 'The social functions of science', *Science and Public Policy*, Vol. 9 (5).

Reskin, B. (1978) 'Sex Discrimination and Social Organisation of Science', In J. Gaston (ed), *Sociology of Science* (San Francisco: Jossey Bass).

Richards, E. and J. Schuster (1989) 'The Feminine Method as Myth and Accounting Resource', *Social Studies of Science*, Vol. 19, pp.697–720.

Rose, H. and S. (eds) (1976) *The Political Economy of Science* (London: Macmillan).

Rose, H. (1987) 'Hand, Brain and Heart: A Feminist Epistemology for the Natural Sciences', in S. Harding and J. F. O'Barr (eds), *Sex and Scientific Inquiry* (Chicago: University of Chicago Press).

Royal Society (1985) *The Public Understanding of Science* London).

Rupke, N. (1987) (ed.), *Vivisection in Historical Perspective* (London: Croom Helm).

Schiebinger, L. (1987) 'The History and Philosophy of Women In Science: A Review Essay', in S. Harding and J. F. O'Barr (eds), *Sex and Scientific Inquiry* (Chicago: University of Chicago Press).

Schumacher, E. F. (1973) *Small Is Beautiful* (Harmondsworth: Penguin).

Stanworth, M. (1987) (ed.), *Reproductive Technologies* (Cambridge: Polity Press).

Werskey, P. G. (1978) *The Visible College* (London: Allen Lane).

Wynne, B. (1988) 'Unruly Technology: Practical Rules, Impractical Discourses and Public Understanding', *Social Studies of Science*, Vol. 18, pp.146–67.

Young, R. (1977) 'Science is social relations', *Radical Science Journal*, Vol. 5.

Yoxen, E. (1988) *Public Concern and the Steering of Science*, SPSG Concept Paper no. 5 (London).

7 Conclusion

Collingridge, D. and C. Reeve (1986) *Science Speaks to Power: The Role of Experts in Policy Making* (London: Frances Pinter).

Cozzens, S. E. (1986) 'Funding and Knowledge Growth', *Social Studies of Science*, Vol. 16, pp.9–21.

Fujimura, J. (1987) 'Constructing Do-able Problems in Cancer Research: Articulating Alignment', *Social Studies of Science*, Vol. 17, pp.257–93.

Harding, S. (1986) *The Science Question in Feminism* (Milton Keynes: Open University Press).

Hoch, P. (1990) 'Institutional Mobility and Science Policy', Paper presented at 'Policies and publics for science and technology' Conference, Science Museum, London.

Hughes, T. (1983) *Networks of Power* (Baltimore: Johns Hopkins University Press).

Hull, F. (1988) 'Inventions from R&D: Organisational Designs for Efficient Research Performance', *Sociology*, Vol. 22 (3), pp.393–415.

Schwarz, M. and M. Thompson (1990) *Divided We Stand – Redefining Politics, Technology and Social Choice* (Brighton: Harvester).

Vergragt, P. (1988) 'The Social Shaping of Industrial Innovations', *Social Studies of Science*, Vol. 18, pp.483–513.

Wright, S. (1986) 'Molecular Biology or Molecular Politics? The Production of Scientific Consensus on the Hazards of Recombinant DNA', *Social Studies of Science*, Vol. 16, pp. 593–620.

Wynne, B. *et al.* (1990) 'Frameworks for Understanding Public Interpretations of Science and Technology', Paper presented at 'Policies and Publics for Science and Technology' Conference Science Museum, London.

Yearley, S. (1988) *The Occupational Culture, Organisation and Exploitability of Scientific Work*, SPSG Concept Paper no. 6 (London).

Index